THE SECRET SCIENCE OF WINNING LOTTERIES, SWEEPSTAKES AND CONTESTS

Laws, Strategies, Formulas and Statistics

$\forall i, x_i \in S_i, x_i \neq x_i^* : f_i(x_i, x^*\_i) \geq f_i(x_i, x^*\_i)$

$i, S = S_1 \times S_2 \dots \times S_n$

$f = (f_1(x), \dots, f_n(x))$

CHARLES JEROME WARE

Outskirts Press, Inc.
Denver, Colorado

The Secret Science of Winning Lotteries, Sweepstakes and Contests
Laws, Strategies, Formulas and Statistics
All Rights Reserved.
Copyright © 2012 Charles Jerome Ware
v1.0

Cover Photo © 2012 JupiterImages Corporation. All rights reserved - used with permission.

Outskirts Press, Inc.
http://www.outskirtspress.com

ISBN: 978-1-4327-9388-3

Outskirts Press and the "OP" logo are trademarks belonging to Outskirts Press, Inc.

PRINTED IN THE UNITED STATES OF AMERICA

OTHER PUBLISHED BOOKS BY THE AUTHOR:

- "Understanding The Law: A Primer", by Attorney Charles Jerome Ware.
- "The Immigration Paradox: 15 Tips for Winning Immigration Cases", by Charles Jerome Ware, Former United States Immigration Judge.
- "Quince (15) Consejos Para Ganar Casos De Inmigracion", por Charles Jerome Ware, Ex Juez de Inmigracon de Estados Unidos.
- "Legal Consumer Tips and Secrets: Avoiding Debtors' Prison in the United States", by Charles Jerome Ware, Former Special Counsel to the Chairman of the United States Federal Trade Commission.

INTRODUCTION

Did you hear about the new million dollar sweepstakes?

The winner gets one dollar a year for a million years.

I have been entering and winning sweepstakes and contests since I was about eight years of age in the small Southern town of Anniston, Alabama. It's fun.

History has shown me that when I have the time, and take the time, to focus on a sweepstakes or contest as I am capable of doing, I have a good chance of winning it.

I have won a lot of sweepstakes and contests, and a lot of "big" prizes. My wins and prizes over the years have included cars, cash, credit and debit cards, televisions, vacation trips, radios, stereos, gift certificates, toys, furniture, appliances and many more. It's been fun.

At age eight back home in Anniston, Alabama, my first win was a Japanese transistor radio. Wow… what fun! In between then and now, my wins have been too numerous to mention. My very latest win was $10,000.00. That was, yes, fun.

I have over time undertaken the project of doing an indepth analysis and evaluation of lotteries. Using what I have learned thus far, and continue to learn about lotteries, I intend to win them too. Why not? It's fun.

My considerable experience has taught me that four major components of winning sweepstakes and contests are patience, persistence, poise and a positive attitude (the "4 P's").

My ongoing and indepth study of lotteries is teaching me the value of statistical analysis and cognitive flexibility learning.

In all of these areas of sweepstakes, contests, and lotteries, luck is certainly important. However, I have discovered that luck can also be enhanced, finessed, and even "made" for all three.

This book proves there is a science of winning lotteries, sweepstakes and contests.

Thanks to my manuscriptist, James Jager, a Shakespearean actor and teacher, for his tremendous word processing skills; as well as my confidential and public sources of information for preparation of this book. Nothing in this book is intended to provide legal advice or consultation to the reader.

I hope and trust this book will help you as much as the experience of writing it has aided me in becoming even better at winning sweepstakes, contests and, now, lotteries.

In dedication to "sweepstakers", "contesters", and "lottericians" everywhere.

TABLE OF CONTENTS

CHAPTER 1

THE DIFFERENCES BETWEEN LOTTER-IES, SWEEPSTAKES AND CONTESTS

Once when I was younger I felt so lucky that I called the Lottery Office to ask for an advance.
-Greg Schiller [writing.gather.com]

Lotteries

A. Lotteries are prize drawings in which, generally, people must pay money to buy a chance to win. The "win" or prize is frequently cash. Lotteries are usually well-regulated by the government, and now-a-days are generally only legal when they are run by the government [http://contests.about.com/od].

A *lottery* is a form of *gambling* which involves the drawing of lots for a prize. Another word for lots is tickets. The usual goal of lotteries is to raise money.

Lotteries come in many forms. For instance, a lottery prize can be a set or fixed amount of cash or other prizes in which the risks fall on the organizer if insufficient lots or tickets are sold. Or, more commonly for government-backed lotteries, the prize fund can be a fixed percentage of the gross revenues or receipts.

A popular smaller-group (usually non-profit) fundraising format of the lottery is the "50-50" draw, in which organizers promise that the cash prize will be 50% of the gross revenue or gross receipts. Recently, many lotteries allow purchasers of lots or tickets to choose the number on the lottery ticket, resulting in the possibility of multiple winners. The best examples of these, of course, are the state lotteries and multi-jurisdictional "power ball"-type lotteries.

Except for the numbers, the Lottery is neutral.

The lottery is about numbers. The numbers control whether you win the lottery or not. No individual's personality, character, religion, education, heritage, race, ethnicity, adult age or background is considered by the lottery as to whether one wins or not.

The Ex-Convict

The lottery does not discriminate, <u>except</u> in favor of those who have the right numbers; for whatever reason. For example, consider the case of the ex-convict who won the $57 million Mega Millions jackpot in a state lottery not too long ago.

It is reported that when this winner won the state's lottery, he then had the distinction of being featured on three of the state's agency websites at the same time:

- the state's lucky lottery winner website;
- the state's public sex offender registry; and

- the state's department of corrections felony offender profile website.

This new multi-million dollar lottery winner had been convicted and served time in prison for 3 felony convictions, including a recent jail sentence for attempted sexual assault.

The lottery does not care about who you are. It only cares about the numbers. Get the numbers right and you get the money.

The "Vivacious Octogenarian"

As another example, take the case of the "Vivacious Octogenarian": 81-year old Louise White, of Newport, Rhode Island, an African-American.

Ms. White, in March 2012, became the winner of the third-largest Rhode Island Powerball Lottery jackpot in the history of the game: a whopping $336.4 million! That's $336,000,000.00!

Her winning Rhode Island Powerball ticket, which cost $3, was the only ticket that matched the winning number combination of:

1-10-37-52-57 and Powerball 11.

She purchased her winning ticket at a Stop & Shop supermarket in Newport.

For tax, accounting, financial and other reasons, Ms. White had her lawyers create a trust, "The Rainbow Sherbert Trust", and accepted her win on its behalf. She named the trust after the ice cream dessert she purchased while buying the winning lottery ticket. Appropriately.

The "Vivacious Octogenarian" Louise White's monstrous win was the sixth largest among all lotteries in the United States at that time. Her reported chances of winning the Powerball were 1 in 175 million.

[usnews.msnbc.msn.com/news/2012/03/06, "$336 million: Rhode Island woman, 81, wins Powerball"]

Good statistical analysis and cognitive flexibility learning, in combination with luck *can* help you *win* lotteries.

[see, "Lucky Woman Who Won Lottery Four Times Outed As Stanford University Statistics Ph D", The Daily Mail, August 7, 2011, www.dailymail.co.uk/news/article-2023514].

However, for the general public, the purchase of lottery tickets ordinarily cannot be accounted for by decision models based upon expected value maximization. **In other words, unless you know precisely what you are doing statistically and cognitively, purchasing a lottery ticket isn't worth the money you pay for it.**

The not-unrealistic joke is that the chance for the ordinary person of winning the average state lottery is something like sixty million to one – the same odds of being hit by lightning ten times in a row.

The reason is that generally the odds of winning the lottery (especially the large or mega-lotteries) are statistically too far against the purchaser. In other words – believe it or not, but for the moment, indulge me – the $1.00 or so you pay for a 100 million-dollar lottery ticket will cost you more than (relatively speaking) the unrealistically expected 100 million-dollar gain. Therefore, without strategic statistical analysis and cognitive flexibility learning to support him or her, the person attempting to maximize the expected value of the lottery probably should not buy lottery tickets.

No Numbers Personality

Many lottery players make the mistake of choosing numbers as if the numbers have personalities. For example, they play their birthday numbers, age numbers, social security numbers, license plates numbers and other so-called numbers associated with events, etc., in their lives. This approach rarely works [see, Chapter VI, B (10), infra].

It is laughable, and is in fact a joke, to think that numbers have personalities:

153 and 641 were dating, and 153 said to 641, "I love you 641. I love how you're a sum of fourth powers. I love how you're a divisor of a Fermat number."

641 replied, "I, uh, I love you too, 153."

"Why do you love me, 641?" asked 153.

"Well, I, uh, I love how you're the sum of the cubes of your digits."

"You don't love me!" cried 153. "You just like my base ten representation!" ☺

[Jim Ferry, corklebath#NoSpam.hotmail.com]

If you find the above story to be humorous, you get the point. Congratulations!

For lottery purposes, numbers have neither personality nor sentimentality. Even adding up and playing the combined ages of your beautiful and loveable children is no more of a lucky approach:

> A gambler and a mathematician went to their local mini-mart to buy "Pick 3" lottery tickets. The mathematician wanted to review the previous winners list to see what the prior winning numbers were, but the gambler insisted that he had a secret algorithm.

> The gambler picked his two children's ages (2 and 5) and added them up for his third "Pick 3" number, resulting in "2-5-7". He did not win.

> The mathematician studied the previous winning ("hot") numbers for the month and won: "3-6-9".

Statistical analysis and cognitive flexible learning win lotteries, not personality and sentimentality.

[see, "Lottery Winners: The Myth and Reality", H. Roy Kaplan, Journal of Gambling Studies 3 (1987; The Lottery Encyclopedia, Ron Shelley, Byron Publishers (1989); "On the Lottery Problem", Z. Furedi, G.J. Szekely, and Z. Zubor, Journal of Combination Designs, Wiley Pub. (1996); "The Relativity of Luck: About the Life of Lottery Millionaires" (Die Relativitats theorie des Glucks. Uber da Leben von Lottomillionaren), Christoph Lau and Ludwig Framer, Herbolzhein: Centaurus (2005)]

The Compulsive Sweepstaker

Filer was a compulsive sweepstaker. He would enter any and all sweepstakes, as well as numerous contests.

When Filer was down to his very last dollar, he went to his best friend, James, asked for $1,000, saying he needed the money because he and his family had no food, owed back rent, needed to get clothes for the children to go to school, and his car had been repossessed.

"Can you help me out?" he asked his friend.

So, Filer's best friend gave him $2,000 to help get him back on his feet, but insisted on one condition --- that Filer does not use the money for any sweepstakes or gambling.

To that demand, Filer's response was, "Oh, I have money put away for that."

Sweepstakes

B. _Sweepstakes_ are prize giveaways in which the winners are essentially chosen by luck, with some notable exceptions as outlined in this book.

I know a lot about sweepstakes.

With persistence and research your "luck" can be enhanced, and even made, with sweepstakes. Sweepstakes prizes can cover the gamut from t-shirts and stickers; to credit cards and cash, from vacations to cars, to houses. I have won all of the above with the exception of a house. I think I am due, though.

Sweepstakes or *sweeps* are generally associated with marketing promotions, and are designed or geared to generate excitement and enthusiasm among potential customers or clients by providing incentives for them to submit free entries into drawings of chance. Since sweepstakes are considered games of chance, it is illegal to charge an entry fee.

These drawings of chance are tied to products or services of the sponsors.

Sweepstakes are *not* contests of *skill*. Remember: they are considered drawings of *chance*.

Probably two of the most difficult sweepstakes in the United States to win are the Publishers Clearing House and the Reader's Digest sweepstakes; for the following reasons:

1. The prizes are large (cash, etc.) and thus attract a lot of entrants;
2. The sweepstakes are marketed to the public extremely aggressively (the Internet, television ads, repeated mass mailings, etc.);
3. The entry process is extremely involved and difficult, and long;
4. The entry process can be tricky and misleading, particularly to the more vulnerable of society such as the elderly; and
5. These sweepstakes, and some other large sweepstakes, are actually marketed as if they require some skill in order to win. Which they do not. They are still based on chance (luck).

Sweepstakes and lotteries are firmly regulated by both state and federal law.

Sweepstakes typically have 2 of the three characteristics of a lottery: chance and a prize. Consequently, to avoid classification as a lottery, a sweepstakes promotion must not involve *consideration* (i.e., the inducement to a contract, or something of value given in return for a performance or a promise of performance by another).

Technically, probably, characterization of a sweepstakes as an illegal lottery could be avoided by eliminating the element of a prize. But, this would be unrealistic. Who wants to take the time and effort to enter a sweepstakes with no prize?

For example, so-called "amusement" gambling such as the playing of pinball and many video games is not generally considered gambling, in that the only prize, *per se*, awarded is more playing time. However, for the legal purists, even this proposition may come under challenge (see, United States v. Sixteen Electronic Gambling Devices, 603 F. Supp. 32 (D. Haw. 1984): where the court concluded that meters to record credits, plus knockoff switches to reset counts for the new player, indicated use of gambling).
[Law Dictionary, Baron's Legal Guide, 3rd Edition (1991); "An Introduction To Sweepstakes And Contests Law", Steven C. Bennett, The Practical Lawyer (August 2007)]

Both the Publishers Clearinghouse and the Reader's Digest Sweepstakes operate by strong-arming hopeful entrants to purchase magazine subscriptions by placing stickers on "contest" entry cardstock while promising multi-million (annuity) winners who will be "announced on TV".
[see, "Lotteries and Sweepstakes", Cecil L'Estrange Ewen, B. Blom Publisher (1972); U.S. Federal Trade Commis-

sion, www.ftc.gov/opa/2006/10; U.S. Federal Trade Commission, "Telemarketing and Telephone Services: Prizes & Sweepstakes",
www.ftc.gove/bcp/menus/consumer/phone/prizes.shtm;
U.S. FTC, "New Spin On Sweepstakes Scams", www.ftc.gov/opa/2005/11/sweeps.shtm]

Contests

C. _Contests_ generally involve some _skill_. Winners are selected in contests based upon some merit, unlike sweepstakes which are based on chance, and lotteries which are considered a form of gambling.

In contests, for example, the person judged to have the best essay, the best photograph, the chosen recipe, the most moving poem, and so forth, wins. Contests are usually allowed to charge an entry fee.

Many people mistakenly use the terms "sweepstakes" and "contests" interchangeably. Technically, as we have stated, the two are different. Contests are giveaways that have some level of skill involved in them; sweepstakes are games of chance or "luck".

Non-entry fee drawings, raffles, etc. are sweepstakes. Checkers, basketball, football, essay and photo shoots, etc., are contests.
[see, contests.about.com/od/sweepstakesandcontests; www.contestcen.com/swcon.htm]

Probability

D. Probability

When playing the lottery, entering sweepstakes, as well as competing in contests, the thrust of your strategy should be to increase (↑) your *probability* of winning.

Probability starts with *logic*, and logic refers to *reason*.

Probability is defined as the rapport of the favorable cases over total cases, or calculated as:

$$P = n/N$$

Be reasonable and logical when playing lotteries, sweepstakes, and contests, and you will increase (↑) your probabilities of winning!

[Also see, "Example for Creating 'Pick-3' (and even 'Pick-4') Number Lotto Algorithms Using Logic", Chapter VI, infra, as well as "Winning Pick-3 and Pick-4 Lottery Games", Chapter V, infra].

Ahhh... Probability!

A statistics and mathematics professor was in the process of traveling to a conference by plane. When he passed through the mandatory airport security check, the security screeners discovered a bomb in his carry-on luggage.

He was, of course, carried off immediately for interrogation:

"I don't understand it!" the interrogating officer yelled. "You are an intelligent and accomplished professional, a caring and loving family man, and a pillar of your community. Why would you risk all of that and destroy everything you have stood for by blowing up an airplane and even killing yourself and others!?"

"I'm sorry," the professor interrupted him. "I never intended nor wanted to blow up the plane."

"Ok, so, for what reason did you attempt to bring a bomb aboard the plane?" the security officer asked.

"Well, let me explain," the professor continued. "Statistics show that the probability of a bomb being on an airplane is 1 in 10,000. That's quite high if you think about it. In fact, it's so high that I couldn't have any peace of mind on a flight."

"But what does that have to do with you bringing your own bomb on board a plane?" the security officer pressed.

"Well, you see, since the probability of one bomb being on my plane is 1 in 10,000, the chance that there are two bombs on the plane is 1 in 1,000,000. Therefore, if I already bring one, and the chance of there being a second is 1 in a million, I and the other passengers are much safer when I bring my own bomb."

[www.jokebudda.com/probability]

Remember: Probabilities do not guarantee winnings; they just propose a likelihood of winning.

CHAPTER 2
WHAT IS GAMBLING?

A minister, an accountant, and a lawyer are playing poker when the police raid the game. Turning first to the minister, the police office asks:

"Reverend, were you gambling?"

The minister lifts his head, turns his eyes to Heaven, and murmurs, "Lord, forgive me for what I am about to do." He then turns to the officer and says, "No, officer, it was just a social game with my friends."

The police officer then quizzes the accountant: "Mr. Ledger, were you gambling?"

Again, following the lead of his friend, the minister, the accountant replies, "No, officer, I was not gambling."

Finally, approaching the lawyer, the officer again asks: "Counselor, were you gambling?"

Shrugging his shoulders, the lawyer answers: "So, with whom would I be gambling?"

[excerpted, www.onlinecasinomansion.com/gamblingjokes.html]

At its core, gambling is "entertainment" or "recreation".

Generally, *gambling* is defined as the practice of playing games of chance or betting in the hope of winning money. Some other names for gambling are betting, gaming, and bookmaking.
[see, Encarta ® World English Dictionary (North American Edition), 2009]

More technically, *gambling* is defined as the wagering of money or something of material value (referred to as "the stakes") on an event with an uncertain outcome. The primary intent or goal of gambling is to win additional money and/or material goods. Ordinarily, the outcome of the wager is evident within a relatively short period of time.
[see, "Gambling", Berel Wein, www.torah.org/features/secondlook/gambling.html (Retrieved July 20, 2010); Gambling Law US, www.gambling-law-us.com; "You Bet", The Economist, July 8, 2010; "Gambling Had Role In Religious History", Rich Barlow, The Boston Globe, 12/01/2007, www.boston.com/news/local/articles]

Gambling in the United States

"Gambling is inevitable. No matter what is said or done by advocates or opponents of gambling in all its various forms, it is an activity that is practiced, or tacitly endorsed,

by a substantial group of
Americans."

-- Commission on the Review
of National Policy toward
Gambling, 1976, p.1.

Legal gambling activities include *state lotteries*; pari-mutuel betting on horses, greyhounds, and jai-alai; sports book-making; card games; keno; blackjack machines; and video roulette machines. Not all of these are legal in all places. Note, however, that sweepstakes and contests are not included.

Gambling is very popular in the United States. According to the American Gaming Association, legal gambling revenues in the United States for 2007 were as follows:

LEGAL GAMBLING REVENUES IN THE UNITED STATES (2007)

---

(1) Legal Bookmaking: → $168.8 million
(2) Card Rooms: → $1.18 billion
(3) Charitable Games and Bingo: → $2.22 billion
(4) Pari-Mutuel Wagering: → $3.50 billion
(5) **LOTTERIES:** → $24.78 billion
(6) Indian Casinos: → $26.02 billion
(7) Commercial Casinos: → $34.41 billion

GRAND TOTAL: $92.27 billion

[see, "Gaming Revenues (in the United States) for 2007", Industry Information: Fact Sheets: Statistics, www.americangaming.org/Industry/factsheets, American Gaming Association]

The Odds of Winning in Gambling

"Knew that we ventured on such dangerous seas
That if we wrought out life 'twas ten to one"

- William Shakespeare, Henry IV, Part II, Act 1, Scene 1 lines 181-2.

Gambling universally invokes the hopeful notion of getting rich quick. It rarely happens.

This idea of getting rich quick has helped casinos and lotteries make billions of dollars every year. Realistically, though, the odds of winning a substantial amount of money *in gambling* without losing a lot of money in the process are remote, or very low.

Casinos essentially have what I call licenses to steal. That is, the reason they make such enormous amounts of money is because the house (the casino) has an enormous advantage over the customer (player).

No matter what game the casino allows you to play, the probability of winning each time is in the casino's favor. The player (customer) is always at a considerable disadvantage when gambling at a casino.

The *"Professional Gambler"*

It is because the odds and probabilities of someone winning on a regular or consistent basis are so remote that the concept of a person being a "professional" gambler (making a living at gambling) is not practical nor realistic.

I have observed that, at best, the most fortunate (or lucky) gambler will **lose at least 52% of the time**. In casinos and gambling "houses" all over, the person or player who wins more than he or she loses is considered a "cheat" or a "card counter" (in poker or blackjack card games), and is therefore banned from participating.

Remember this: When you gamble, you are paying your money for "entertainment".

Odds and Probabilities

Far too many "gamblers" do not understand what probabilities and odds really mean. Consider the following scenario:

Assume that a gambler's odds of winning a particular prize in a drawing or raffle are 1 in 10. This means that if only 10 raffle tickets are sold and put "in the hat" so to speak, the hopeful gambler would have one ticket in the drawing.

However, if the drawing or raffle sold 100 tickets and drew for 10 prizes, the gambler is still not guaranteed to win, even though the <u>odds</u> suggest he or she will win.

Odds can be misleading to gamblers, therefore it is helpful for the player to understand the mathematics behind <u>probabilities</u>.

Probabilities do not guarantee winnings. Probabilities simply propose a likelihood of winning. [mathcentral.uregina.ca/beyond/articles/gambling/odds.html]

The Concept of "Independent Events"

The Concept of "Independent Events" helps explain the odds of winning in gambling. It is frequently misunderstood by gamblers. Independent events is a probability term meaning that past events actually have no influence on future outcomes.

In other words, in "gambling" history does not matter.

Take, for example, the flipping of a coin four consecutive times. Under the concept of independent events, the probability of getting four heads in this scenario is as follows:

$$(\tfrac{1}{2})(\tfrac{1}{2})(\tfrac{1}{2})(\tfrac{1}{2}) = \tfrac{1}{16}$$

The reason for this is because the probability of flipping a head if you flip a coin is $\tfrac{1}{2}$ (an independent event). Therefore, no matter how many times you flip a coin, the probability of getting a head remains the same: $\tfrac{1}{2}$.

The problem that many gamblers have, however, is that they believe or hope that the first three flips (for example) will somehow influence the fourth flip. It does not. That is a misunderstanding. The probability remains the same for each and every flip of the coin: $\tfrac{1}{2}$.

In "gambling" history does not matter.

This same misunderstanding also occurs frequently with people who are picking numbers for the lottery. Simply picking the same numbers faithfully or religiously every time you play a lottery does not guarantee that these numbers will eventually be picked.

Further, when playing the lottery, if a player (for example) picked 40 as one of their numbers and one of the winning lottery numbers was, say, 39, it does not mean or even imply that the player was *close* to winning that lottery. This, again, is an example of the player's lack of knowledge or misunderstanding of the "concept of independent events".

[mathcentral.uregina.ca/beyond/articles/gambling/odd, Natasha Glydon, Canadian Mathematical Society, University of Regina, Imperial Oil Foundation]

Before you gamble... know the odds!

CHAPTER 3
THE "4 P'S"

My research experience and observations have taught me that four major components of winning lotteries, sweepstakes, and contests are (a) patience, (b) persistence, (c) poise and (d) a positive attitude (the "4 P's"):

A. <u>Patience.</u>

I want patience… and I want it NOW!!!

Patient Customer Service

The exhausted clerk had pulled down blanket after blanket from the shelf, but still the woman customer was not satisfied.

"There is one more blanket left," said the clerk. "Do you care to see it?"

"I'm not going to buy one today," said the woman. "I have only been looking for a friend."

"Well," said the clerk, "I'll take the last one down if you think your friend might be in it."
[www.jokebuddha.com/Patience]

Wittingly, "patience" could be defined as a minor form of despair, disguised as a virtue.

From the average dictionary, "patience" is usually defined as the capacity for waiting: the ability to endure waiting, delay, or provocation without becoming annoyed or upset, or to persevere calmly when faced with difficulties [www.bing.com/Dictionary].

Since "patience is the best remedy for every trouble" [Titus Maccius Plautus, 254 BC-184 BC, Rudens], and since "patience is the companion of wisdom" [Saint Augustine, 354 AD-340 AD], I strongly recommend the reader learn the art of patience.

Patience can help you achieve your goals.

"Patience" allows us to apply discipline to our thoughts when we become anxious over the outcome of a goal. "Impatience" breeds anxiety, fear, discouragement and failure. "Patience" creates confidence, decisiveness, and a rational outlook, which eventually leads to success [Brian Adams, author].

With respect to lotteries, sweepstakes, and contests, patience (or forbearing) enters into play as the state of endurance for the player or person under the difficult circumstances of waiting in anticipation of winning, frustration over choices not made, and so forth.

The science of patience (or "self-control") involves evolutionary psychology and cognitive neuroscience studies and observations of decision-making of problems. For example, whether a person (or other animal) chooses a small reward in the short term, or a more valuable reward in the long term (which involves more patience or self control), goes directly to the issue of that person's "self-control".

Scientifically, it is established that, when given a choice ordinarily, humans (and other animals) are inclined to favor or choose short-term rewards over long-term rewards. This is true despite the frequently greater benefits associated with long-term rewards. ["The Ecology and Evolution of Patience in Two New World Monkeys", www.pubmedcentral.nig.gov; ""Patience", Reuven Firestone, www.huc.edu/chronicle/60/2008/06/12]

In sum, the patient lotterician, sweepstaker and contester will apply discipline when playing and, thus, create confidence, decisiveness, and a rational outlook towards the games; which will eventually lead to success.

"A handful of patience is worth more than a bushel of brains"
- Dutch Proverb

Five Steps and 12 Sub-Steps to Becoming More Patient

1. Find Out What Makes You Impatient.
    Try to understand why you are rushing, or are anxious.
    Nail down or pinpoint the things (actions, etc.) that frequently influence you to lose patience.

Look for patterns in your impatience (patterns in what makes you impatient).

2. Write down (keep a journal of) your incidents of impatience. When you feel impatient, record it.

3. Work to overcome your impatience.
Fight or battle your impatience. Try to change your attitude about life. Try to relax. Take some deep breaths when you feel impatient.
If you find (not feel) you cannot do anything about your impatience, let go. Do not angst or worry about it.

4. Look at the Big Picture.
Do not become impatient with yourself. Remind yourself that improvements take time.
Stay focused on what really matters most in your life. Usually it is not the issue or thing that makes you impatient.
Stay focused on what you want. Follow the advice in this book. You will get what you want, eventually.
Always keep a positive outlook on life.

5. Pause. Take a break. Step back and look at yourself and at life in general.
Accept the twists and turns in life gracefully. Be poised. Expect the unexpected.
Slack off. Sit quietly and think. Give yourself a break. Enhance your calm.

Remember: *There are two cardinal sins from which all others spring. Impatience is one. The other is laziness.* --- Franz Kafka (1883-1924).

B. Persistence.

Inspiring Persistence

Soichiro Honda was a simple Japanese mechanic who worked mainly out of his modest home. He was audacious enough to attempt to design and build a new type of piston engine to improve the performance of Japanese cars, despite a severe shortage of money and other resources.

In order to make some money to live he offered his designs to Japanese automobile manufacturer Toyota. Toyota turned him down without even the dignity of meeting him.

But, Soichiro Honda did not give up.

He kept working on and improving his designs, and he repeated his efforts to meet with Toyota engineers. Finally, he was permitted a meeting in which his designs were totally dismissed, and even ridiculed, by the engineers.

But, Soichiro Honda did not give up.

Constantly learning and improving his piston designs (using cognitive flexibility learning), he eventually secured a very small order from Toyota to supply pistons to the company.

He invested all of his resources --- his savings, materials, knowledge, experience, and hardwork – into this new contract with Toyota and he constructed a manufacturing plant to produce the pistons.

Suddenly an earthquake occurred in Japan that totally destroyed his plant.

But, Soichiro Honda did not give up.

He started all over, built a new plant from scratch and was set to reopen the following week when World War II broke out. Bombs destroyed his new factory, and, of course, devastated the country of Japan. Again, Honda was put out of business.

But, Soichiro Honda did not give up.

Still, having lost his factory, resources, and even his friends again, he persisted and persevered. He began to construct his factory for the third time; and went on to manufacture his own cars.

Because of his rational persistence (and his use of cognitive flexibility learning), mechanic Soichiro Honda's car company, Honda, is one of the world's largest and most respected automobile manufacturers in the world.

The moral of this story is: *Rational persistence (and cognitive flexibility thinking) equals success.*

"Persistence" is the continuance of an effect after the cause of the effect has stopped; continued effort or existence; the property of a continuous and connected period of time.

[Collins English Dictionary, Harper Collins (1991, 1994, 1998, 200, 2003); www.thefreedictionary.com/persistence].

"Persistence" in gambling is futile. Overall, over a period of time at least, you will lose more money than you win no matter how persistent you are.

"Persistence" in lotteries, sweepstakes and contests, however, can be helpful to you in winning; particularly when combined with the other tips, principles, and secrets in this book; including cognitive flexibility learning.

The Science of Persistence

There is, in fact, a science of persistence.

For instance, in psychology "persistence" (PS) is a personality trait. It is measured in the Temperament and Character Inventory. The subsets/substeps of "persistence" (PS) consist of:

      (1) Eagerness of effort (PS1)
      (2) Hardened work (PS2)
      (3) Ambition (PS3), and
      (4) Perfectionism (PS4)
            Or
    $PS = PS1 + PS2 + PS3 + PS4.$

Further, in computer science "persistence" refers to the "characteristic of state" that outlives the process or program that created it; or, in layman terms, the attitude of continuing with the task despite obstacles encountered [en.wikipedia.org/persistence].

In mathematics, the "persistence" of a number refers to the number of times one must apply a given operation to an integer before reaching a fixed point, i.e., until further application does not change the number any more; or, in other words, the number of times (amount of effort) a person must play a game (a certain way) in order to win the game [en.wikipedia.org/wiki/Persistanceofanumber].

If you understand this, mathematically you are ahead of the curve. ☺

C. Poise (Do not panic).

Poise is the ability to continue speaking entertainingly as the other fellow picks up the restaurant check. Oops! ☺

Poise in the Caribbean

A monster hurricane blew across the Caribbean. Soon, the expensive yacht was swamped by the high waves and it sunk. Only two survivors remained: the steward, Jack, and the yacht's owner, Dr. Smith; both of whom managed to swim safely to the nearest deserted island.

After reaching the island, the steward was crying uncontrollably and was extremely upset that he and Dr. Smith may never be found and rescued.

Dr. Smith on the other hand was quite calm and collected, poised even, relaxing in the shade against a tree.

"Dr. Smith, Dr. Smith, how can you be so calm?" cried steward Jack. "We're going to die here on this deserted island! We will never be discovered and rescued!"

"Sit down young man and listen to me," began Dr. Smith with considerable confidence. "Five years ago, I gave the United Way $500,000 and another $500,000 to my church. Four years ago, I did it again and donated the same amounts. And, three years ago, I did extremely well in the stock market, so I contributed $750,000 to each again. Last year, my business was so good --- plus I won the lottery --- that I gave each charity a million dollars", stated Dr. Smith.

"So what?" shouted the frantic Jack.

"Well, it's now time again for their annual fundraising drives, and I just know they're going to find me!" smiled the poised Dr. Smith.
[www.jokebuddha.com/calm]

Most dictionaries define "poise" something like this:
To carry or hold equilibrium; balance; a state of balance or equilibrium; stability; freedom from affectation or embarrassment; composure; unflappable; collected; self-composure; dignity.

I personally view people who I see as having "poise" as being calm, composed, and collected.

The Science of Poise

The science of poise can be seen in the fields of chemistry, medicine, and biochemistry.

The Oxford Dictionary of Chemistry defines "poise" as:
A unit of viscosity equal to the tangential force in dynes per square centimeter required to maintain a difference in velocity of one centimeter per second

between two parallel planes of a fluid separated by one centimeter. 1 poise is equal 10^{-1}Nsm^{-2}.

In other words, "Poise" in chemistry represents, again, stability and calmness. Balance.

[see, "Viscosity of Liquids," CRC Handbook of Chemistry and Physics, 89th Edition (Internet Version 2009), David R. Lide, Ed., CRC Press/Taylor and Francis, Boca Raton, Fl.].

Do not panic. Enhance your calm. Be poised when playing the lottery, entering sweepstakes, and competing in contests.

D. Positive Attitude.

Major Attitude

Former First Lady of the United States, Eleanor Roosevelt, had a wonderful attitude about life. And, she was self-deprecating. Eleanor Roosevelt wittingly said:

"I once had a rose named after me and I was very flattered. But I was not pleased to read its description in the flower catalogue: no good in a bed, but fine up against a wall." [www.seniorresource.com/attitude.htm]

The Science of Attitude

An "attitude", psychologically, is a favorable or unfavorable evaluation of something. Attitudes are generally positive or negative views of a person, place, thing, or event.

["The Neuroscience of Social Decision-Making," Annual Review of Psychology (62): 23-48, January 2011; "How

Do Attitudes Guide Behavior?", R.H. Fazio (1986), The Handbook of Motivation and Cognition Foundations of Social Behavior (pp. 204-243), New York: Guildford Press]

"Attitude" is also very important in flying airplanes. Airplanes have an attitude indicator (AI), also known as a gyro horizon or artificial horizon, which is an instrument used to inform the pilot of the orientation of the aircraft relative to earth. In the years I flew planes, the attitude instrument was one I gave a lot of attention to when flying.

In flying a plane I always remember that **"Performance (of the aircraft) = Attitude + Power", or P = A+P.**

["NTSB Safety Recommendation", National Transportation Safety Board, 07-22-2008; www.wordiq.com/definition/aircraftattitude]

A positive attitude is helpful in so many ways: coping with the daily affairs of life, making it easier to avoid negative thoughts and worry, and generally helping you feel better about living.

A positive attitude brings optimism into a person's life, and if you adopt it as a way of life it will bring constructive changes into your life. A positive attitude makes people happier, brighter and more successful.

Positive people see the bright side of life, are optimistic, and expect the best to happen to them and to others. A positive attitude is a state of mind that is well worth developing and strengthening.

A positive attitude is extremely useful for those seeking success with lotteries, sweepstakes, and contests.

[www.successconsciousness.com/positiveattitude.htm]

Some Specific Attitude Issues To Achieve

There are several specific attitude issues I recommend the reader consider achieving for success in lotteries, sweepstakes, and contests:

(1)   Attitude 1: Take calculated risks in life.
(2)   Attitude 2: Respect your own well-thought out opinions.
(3)   Attitude 3: Be mentally adaptable and flexible (use cognitive flexibility thinking).
(4)   Attitude 4: Take on new attainable challenges and learn new things.
(5)   Attitude 5: Deal with pain, losses, and disappointments, but do not hold on to the suffering they bring.
(6)   Attitude 6: See the glass half-full, not half empty.
(7)   Attitude 7: Take care of your health. Make your good health a priority.
(8)   Attitude 8: Do not accept society's myths as being true about you.
(9)   Attitude 9: Remember that worry is a sin.
(10) Attitude 10: Enhance your calm (poise).

Words of Wisdom about Attitude

"Nothing can stop the man with the right mental attitude from achieving his goal; nothing on earth can help the man with the wrong mental attitude."

- Thomas Jefferson

"The longer I live, the more I realize the impact of attitude on life. Attitude, to me, is more important than facts. It is more important than the past, the education, the money, than circumstances, than failure, than successes, than what other people think or say or do. It is more important than appearance, giftedness or skill. It will make or break a company... a church... a home. The remarkable thing is we have a choice everyday regarding the attitude we will embrace for that day. We cannot change our past... we cannot change the fact that people will act in a certain way. We cannot change the inevitable. The only thing we can do is play on the one string we have, and that is our attitude. I am convinced that life is 10 percent what happens to me and 90 percent how I react to it. And so it is with you... we are in charge of our attitudes."

- Charles Swindoll

"The greatest discovery of my generation is that a human being can alter his life by altering his attitudes of mind."

-William James

There they are, the "4 P's": patience, persistence, poise, and a positive attitude. Practice them for winning success.

CHAPTER 4
LOTTERIES IN THE UNITED STATES

The No-Nonsense Lottery Winner

A no-nonsense elderly woman walked into a bank and said to the male teller at the counter:

"I want to open a damn checking account."

To which the stunned male teller replied:

"I beg your pardon, ma'am. I must have misunderstood you. What did you say?"

"Listen carefully, damn it!" the elderly woman continued, "I said I want to open a damn checking account now... today!"

To which the shaken teller responded:

"I'm very sorry, ma'am, but we do not tolerate that kind of foul language in this bank."

As the woman stubbornly waited for service, the teller left the counter, went over to the experienced bank manager and told her about the unfortunate encounter with the woman customer. Both the male teller and the female bank manager returned promptly to confront the woman.

"Exactly what seems to be the problem here?" asked the manager to the customer.

"There's no friggin problem, dammit!" the customer replied with much irritation, "I just won $100 million in the damn lottery and, up to now, I wanted to open a damn checking account in this damn bank!"

"Oh! I see!" replied the now excited bank manager as she turned with severe displeasure on her face and stared at the teller, "And this damn idiot is giving you a hard time?!"

[Revised from www.lottomania.ch/humour.php]

Very little in life creates a more accepting and positive view of a person than that person winning the lottery, a sweepstakes, or a contest.

At the present time, there are approximately 46 legal lotteries in the United States. These legal lotteries are "run" (operated) by the 46 jurisdictions, including 43 states plus the District of Columbia, Puerto Rico, and the United States Virgin Islands.

In the United States, the lottery in each jurisdiction is subject to the laws of that specific jurisdiction. There is no such thing as a "national lottery" (involving money or direct financial gain).

The "Numbers Racquet"

Prior to creation of "legal" government-sponsored lotteries, many "illegal" lotteries flourished in the U.S. mostly in poor or otherwise marginal neighborhoods ["Playing the Numbers: Gambling in Harlem Between the Wars", Shane White, Stephen Robertson and Graham White, 2010, Harvard University: Cambridge, Mass].

These illegal lotteries, colloquially called "numbers", "numbers game", or "numbers racquet", were an illegal daily lottery in which money was wagered on the appearance of certain numbers in some statistical listing or tabulation published in a daily newspaper, racing form, etc. ["Kings: The True Story of Chicago's Policy Kings and Numbers Racketers; An Informal History", 2003, Nathan Thompson, The Bronzeville Press].

Ordinarily the "numbers racquets" in major cities such as New York and Chicago operated something like this: a bettor simply picked 3 numbers (3 digits only) such as, for example, 1-4-7, in an effort to match those three digits with the winning three digits which would allegedly be drawn randomly the next day.

The bettor (or gambler) would make his bet with a "bookie" (a middle man) at a bar, or some other semi-private place (presumably away from law-enforcement) that served as a "betting parlor".

A "runner" carried the money betted by the gamblers (bettors), along with the "betting slips" (the bettors' 3 digits, and identification usually) between the betting parlors and the numbers headquarters, which was called a "numbers bank" or "policy bank". The word "policy" was borrowed from the insurance industry as a so-called gamble on the future.

["The Mafia Encyclopedia", Facts on File, by Carl Sifakis (2005); "Playing the Numbers: Gambling in Harlem Between the Wars", by Shane White, Stephen Garton, Stephen Robertson and Graham White, Harvard University Press: Cambridge, Massachusetts (2010)].

First "Legal" Lotteries in the United States

The first so-called "modern" government-operated lottery in the United States was established in Puerto Rico in 1934 or, according to some, New Jersey.

New Hampshire followed with its own lottery in 1964.

The most recent legal lottery formed in the U.S. was in Arkansas in 2008.

The first modern "joint" (multiple states) lottery was established in 1985, combining on-line games of 3 lotteries.

In 1988 the "Powerball" Multi-State Lottery was formed with Iowa, Kansas, Missouri, Oregon, Rhode Island, West Virginia and the District of Columbia as charter members. It was designed to create large multi-million dollar jackpots.

In 1996, another joint lottery, "The Big Game" (now called "Mega Millions") was created by 6 other state lotteries as its charter members.

The 1970s introduced instant lottery tickets, better known as "scratch cards", which have become a major source of lottery revenue. Several lotteries have introduced "keno" and/or video lottery terminals. Essentially, keno or video lotteries are slot machines.

Three-digit and four-digit lottery games are also growing in number (so-called "numbers games"). Five-digit and six-digit number games often feature a jackpot. The number 1 reason in the United States for creation of lotteries, allegedly, is to support public education systems.

[www.lexjuris.com/LEXLEX/Leyes2006;www.library.ca.gov/CRB/97/03/chapter3.html]

Lottery Jurisdictions in the United States
(As of May 3rd, 2011)

Jurisdiction	Lottery	Mega Millions	Powerball	Other Joint Games	Top 10 Lotteries (Jackpot Pay-Outs)
Alabama	No	-	-	-	Number 7 in Jackpot Payouts (California)
Alaska	No	-	-	-	
Arizona	Yes	Yes	Yes	0	
Arkansas	Yes	Yes	Yes	1	
California	Yes	Yes	No	0	
Colorado	Yes	Yes	Yes	0	Number 3 in Jackpot Payouts (Florida)
Connecti-	Yes	Yes	Yes	0	
cut	Yes	Yes	Yes	2	
Delaware	Yes	Yes	Yes	1	
District of Columbia	Yes	No	Yes	0	
Florida					
Georgia	Yes	Yes	Yes	1	Number 5 in Jackpot Payouts

					(Georgia)
Hawaii	No	-	-	-	Number 10
Idaho	Yes	Yes	Yes	2	in Jackpot
Illinois	Yes	Yes	Yes	0^	Payouts
					(Illinois)
Indiana	Yes	Yes	Yes	0	Number 2
Iowa	Yes	Yes	Yes	2^	in Jackpot
Kansas	Yes	Yes	Yes	2	Payouts
Kentucky	Yes	Yes	Yes	1	(Massachu-
Louisiana	Yes	Yes	Yes	0	setts)
Maine	Yes	Yes	Yes	5	
Maryland	Yes	Yes	Yes	0	
Massa-chusetts	Yes	Yes	Yes	0	
Michigan	Yes	Yes	Yes	0	Number 9
					in Jackpot
					Payouts
					(Michigan)
Minnesota	Yes	Yes	Yes	1	Number 1
Mississip-	No	-	-	-	in Jackpot
pi	Yes	Yes	Yes	0	Payouts
Missouri	Yes	Yes	Yes	2	(New
Montana	Yes	Yes	Yes	1	York)
Nebraska	No	-	-	-	
Nevada	Yes	Yes	Yes	5	
New					
Hamp-	Yes	Yes	Yes	0	
shire	Yes	Yes	Yes	1	
New Jer-sey	Yes	Yes	Yes	0	
New Mexico					
New York					
North Carolina	Yes	Yes	Yes	0	Number 8
					in Jackpot
North Da-	Yes	Yes	Yes	3	Payouts

kota	Yes	Yes	Yes	0	(Ohio)
Ohio					
Oklahoma	Yes	Yes	Yes	1	Number 6
Oregon	Yes	Yes	Yes	0	in Jackpot
Pennsyl-	Yes	Yes	Yes	0	Payouts
vania	Yes	No	No	0	(Pennsyl-
Puerto Ri-	Yes	Yes	Yes	1	vania)
co	Yes	Yes	Yes	0	
Rhode Is-					
land	Yes	Yes	Yes	2	
South	Yes	Yes	Yes	0	
Carolina	Yes	Yes	Yes	0	
South Da-					
kota					
Tennessee					
Texas					
Utah	No	-	-	-	Number 4
US Virgin	Yes	Yes	Yes	0	in Jackpot
Islands					Payouts
Vermont	Yes	Yes	Yes	5	(Texas)
Virginia	Yes	Yes	Yes	1	
Washing-	Yes	Yes	Yes	0	
ton	Yes	Yes	Yes	2	
West Vir-	Yes	Yes	Yes	0	
ginia	No	-	-	-	
Wisconsin					
Wyoming					

[www.dailyfinance.com/2011/05/03/lottery-states-biggest-jackpot; www.library.ca.gov/CRB/97/03/Chapter13.html]

The Top 10 State Lotteries that Pay the Most in Jackpots (as of May 3rd, 2011)

#1. New York State. Largest single jackpot: $336 million (two-state lottery win shared with California lottery, 2009).

#2. Massachusetts State. Largest single jackpot: $294 million (2004).

#3. Florida State. Largest single jackpot: $189 million (2009).

#4. Texas State. Largest single jackpot: $330 million (four-state shared lottery win, 2007).

#5. Georgia State. Largest single jackpot: $390 million (two-state lottery win shared with New Jersey, 2007).

#6. Pennsylvania State. Largest single jackpot: $213 million (2004).

#7. California State. Largest single jackpot: $315 million (2005).

#8. Ohio State. Largest single jackpot: $270 million (2006).

#9. Michigan State. Largest jackpot: $363 million (two-state lottery win shared with Illinois, 2000).

#10. Illinois State. Largest jackpot: $363 million (two-state lottery win shared with Michigan, 2000).

[www.dailyfinance.com/2011/05/03/lottery-states-biggest-jackpots]

POWERBALL

The American lottery game "Powerball" is sold in 44 jurisdictions as a shared (multi-state) jackpot game. Powerball replaced the original "Lotto*America" multi-state lottery in April 1992.

The coordinating body for Powerball is the Multi-State Lottery Association (MUSL), a non-profit organization formed by an agreement with state lotteries. Prior to January 15, 2012, Powerball's jackpots began at $20 million. With a change in operations, Powerball's minimum jackpot as advertised is $40 million (annuity) with a potential of 9-figure prizes. Powerball's annuity option is paid in 30 graduated installments, or the winner may select the option of cash instead.

Powerball uses a 5/59 (white balls) plus a 1/35 (Powerballs) matrix from which winning numbers are chosen.

[Changes Coming to Powerball...", Multi-State Lottery Association (MSLA), www.powerball.com/pb, 01-22-2012; "Frequently Asked Questions", powerball.com/pb, 04/14/2011]

MEGA MILLIONS

The U.S. multi-jurisdictional lottery game "Mega Millions" actually replaced "The Big Game" lottery in May 2002.

Mega Millions advertised jackpots have started at $12 million, and have paid-out in twenty-six yearly installments, unless of course the cash option is chosen. The jackpot increases each drawing when there is no winner.

Consistent with common practice in American lotteries, the Mega Millions jackpot is advertised as a nominal value of annual installments. The cash value option, when selected by a jackpot winner, pays-out the approximate present value of the annual installments.

Mega Millions presently uses a 5/56 (white balls) +1/46 (the Megal Ball) double matrix to select its winning numbers.

["FAQs", Mega Millions, www.megamillions.com/faqs, 01-12-2009; "Mega Power Lottery", Worldlottery.net, www.worldlottery.net, 12-21-2011]

On October 13, 2009, the Multi-State Lottery Association (MUSL), the coordinator for "Powerball", and the Mega Millions consortium executed an agreement to permit U.S. lotteries to sell both games; thus no longer mandating exclusivity.

CHAPTER 5
HOW TO WIN LOTTERIES 101: COGNITIVE FLEXIBILITY LEARNING

A. Action.

A man named Doubt found himself in deep financial trouble. His business had failed and he needed income to survive.

Doubt was so desperate he decided to ask God for help. He began to pray: "...Please, God, help me. I've lost my business, and if I don't get some money soon I'm going to lose my house as well. Please, God... let me win the lottery."

Lottery night came and somebody else won the lottery.

Once again, Doubt began to pray: "...God, please, why have you forsaken me?!?! I'm hurting here!!! I've lost my business, my home, my car... and I'm getting ready to lose my family! I have never asked you for help before, and I have always been a good and faithful servant to you! Would you please just let me win the lottery just this one time so I can get my family and my life back in order?!?!"

Lottery night came again and Doubt, again, does not win.

Again, Doubt started to pray when... suddenly the Heavens opened, a blinding light appeared and Doubt was immediately confronted with the powerful and authoritative voice of the Almighty God himself:

"Doubt, meet Me halfway on this one, Bud. BUY A TICKET!"

[Modified and excerpted, www.onlinecasinomansion.com/gambling-jokes.html]

Action. In order to win the lottery – you have to buy a ticket and play the lottery.

B. Rational Optimism.

Optimism: a disposition or tendency to look on the more favorable side of events or conditions and to expect the most favorable outcome. [Dictionary.com/reference.com/optimism]

Lotteries, sweepstakes and contests are, of course, games of *optimism.*

However, the people who repeatedly win lotteries, sweepstakes and contests are what I call *rational optimists.* That is, they carefully do their homework on the games, weigh their options, do the necessary statistical analysis as well as cognitive flexibility thinking, and plan --- before playing the games.

Everybody wants good-luck. And, everybody wants to win the lottery.

Don't take just my word for it. Check this out for yourself. Walk into your neighborhood bar and ask the man or woman sitting next to you at the bar, "What would you do if you won the lottery?" Almost immediately, everybody within earshot will join the conversation (and without even being invited to join it). Even the people who claim they don't play the lottery know, or have an opinion about, what they would do if they won it!

The quest for good-luck both demands and invites rational optimism, though.

A major trait of the *rational optimist* is that he or she tends to evaluate, analyze, make adjustments and, when it seems necessary, *change.*

The Effects Of Optimism On Luck

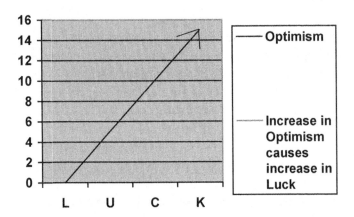

[see, www.psychologytoday.com/articles/201005/make-your-own-luck, Rebecca Webber (May 01, 2010); www.RichardWiseman.com/research/psychologyluck.html; Michael Shermer, "As Luck Would Have It" Scientific American, March 26, 2006; Martin E.P. Seligman, "How to Change Your Mind and Your Life", Learned Optimism, Vintage Pub. (2006); "How to Remain (Reasonably) Optimistic: Scientific Realism and the "Luminiferous Ether", John Worrall, London School of Economics, Philosophy of Science Association (PSA), 1994; "Optimism & Pessimism: Implications for Theory, Research, and Practice", Michael Scheier, Charles Carver, Michael W. Bridges, American Psychological Association (APA), Washington, D.C. (2001)]

C. Make your own luck.

A woman's husband of many years sadly had been slipping in and out of a coma for several months, yet she had faithfully stayed by his bedside every single day.

One day, when he gained consciousness, he feebly motioned his hand for her to come closer to him as he lay weakly in bed.

As she lovingly and carefully sat by him, he whispered softly to her, his eyes full of tears, "You know what? You have been with me all through my many bad times. When I got fired from my job, you were there to support me. When my business failed, you were there too. When I got shot, you were there by my side – on the scene. When we lost the house in foreclosure, you were there with me.

And now, with my dreadful health condition, my illnesses, you are still here by my side. You know what?"

"What dear?" the woman asked gently and warmly to her very ill husband, smiling kindly as her heart filled with love and affection for him.

"I think you're bad luck," he answered.
[Jokes4all.net, 2012]

No one is your bad luck. You make your own luck.

In his articles on luck research, Professor Richard Wiseman, of the University of Hertfordshire in England, proposes 4 "basic principles that lucky people use to create good fortune in their lives":

Principle One: Maximize Chance Opportunities
Lucky people are skilled at creating, noticing and acting upon chance opportunities. They do this in various ways, including networking, adopting a relaxed attitude to life and by being open to new experiences.

Principle Two: Listening to Lucky Hunches
Lucky people make effective decisions by listening to their intuition and gut feelings. In addition, they take steps to actively boost their intuitive abilities by, for example, meditating and clearing their mind of other thoughts.

Principle Three: Expect Good Fortune
Lucky people are certain that the future is going to be full of good fortune. These expectations become self-fulfilling prophecies by helping lucky people persist in the face of failure, and shape their interactions with others in a positive way.

Principle Four: Turn Bad Luck to Good
Lucky people employ various psychological techniques to cope with, and often even thrive upon, the ill fortune that comes their way. For example, they spontaneously imagine how things could have been worse, do not dwell on ill fortune, and take control of the situation.
[http://www.richardwiseman.com/research/psycholo
gyluck.html]

■ Principle 1	25			
□ Principle 2	50			
■ Principle 3	75			
▣ Principle 4	100			

Professor Richard Wiseman's 4 Principles that Lucky People use to create good fortune in their lives.
(The total of these 4 principles is 100%)

In her <u>Psychology Today</u> article, "Make Your Own Luck", May 01, 2010, Rebecca Webber puts forth "Five principles for making the most of life's twists and turns":

(1) <u>See Serendipity Everywhere</u>.

Serendipity: an aptitude or faculty for making desirable discoveries by accident; good fortune.
[<u>Collins English Dictionary</u>, 10th Ed., 2009]

People, for example, who call themselves lucky score higher on the personality factor *extraversion*, and consequently are more likely to have a <u>fortuitous encounter</u> or <u>lucky event</u> because they attract good fortune.

> Extraversion: the act, state, or habit of directing attention toward and obtaining gratification from what is outside the self.
> [Webster's New Collegiate Dictionary, 2010]

(2) Prime Yourself for Chance.

Serendipity smiles upon people who have a more relaxed approach to life. They have clarified their long-term goals but don't worry too much about the details. Rather than aiming to become the top lawyer in the state or the top neurosurgeon at the Johns Hopkins Hospital, they commit to becoming a lawyer who fights justice or a doctor who saves lives. Do not be so quick to condemn a legal lottery opportunity in advance. Check it out first. Analyze... evaluate... think. Be cognitive flexible.

Good news: cognitive flexibility can be cultivated. If you don't have it, you can develop it. Do so.

> Cognitive flexibility: the "flexible" way learners or students assemble and retrieve knowledge from their brains. It involves development of the ability to switch behavioral responses according to the context of the situations. Improvement of the ability to make adjustments.
> ["The Role of Learning Tasks on Attitude Change Using Cognitive Flexibility Hypertext Systems, Journal of Learning Sciences, 13(4)507-526; Godshalk, Harvey and Moller (2004)]

In other words, within the context of winning lotteries, sweepstakes, and contests, cognitive flexibility means learning to change your behavioral patterns based upon your assessment of the facts or situation. Make adjustments to improve your performance.

USING COGNITIVE FLEXIBLITY TO WIN LOTTERIES, SWEEPSTAKES, AND CONTESTS

Mastering the Rules of the Game (<u>Learn</u>)	Following the Rules of the Game (Follow-through)	Playing the Game (Practice)	Evaluating the Results of each Game (Analysis)	Follow the His-tory (Re-sults) of the Game
V	V	V	V	V

COGNITIVE FLEXIBLE LEARNING

Leads to

V

Winning the Lottery, Sweepstakes and Contests!!!

Congratulations! You have won!

[R.J. Spiro, P.J. Feltovich, M.J. Jacobson, & R.L. Coulson, "Knowledge Representation, Content Specification, and the Development of Skill in Situation-Specific Knowledge Assembly: Some Constructivist Issues As They Relate to Cognitive Flexibility Theory and Hypertext", <u>Educational Technology</u>, 31(9), 22-25 (1991); M.H. Roy, Small group communication and performance: do cognitive flexibility and context matter?", <u>Management Decision, 30(4), 323-330 (2001);</u> S.R. Boger-Mehall, "Cognitive Flexibility

Theory: Implications for teaching and teacher education", www.kdassem.dk/didktik/14-16.htm, retrieved July 6, 2007; G. Kearsley, "Cognitive Flexibility Theory", 2007 (R. Spiro, P. Feltovich & R. Coulson), tip.psychology.org/spiro.html]

(3) Go Ahead, Slack Off.

Conscientiousness is no friend to *serendipity*. Therefore, serendipity and conscientiousness cannot occupy the same space at the same time.

Using the definition put forth by Professor Carol Sansone of the Department of Psychology at the University of Utah, and endorsed by Rebecca Webber of Psychology Today (May 01, 2010), *"conscientiousness* means you do what you are supposed to do, and you stick with it."

The problem is, according to Professor Sansone, conscientious people will frequently persist in a task even when there is no good reason to do so! This may explain why it is possible to "try too hard". By rigidly and sternly pouring all of your effort into one approach to win your lottery, or sweepstakes, or contest, the entrant misses out on the unexpected – but more direct – paths to success.

Do not be afraid to do something different. Try a different approach. Make adjustments to improve your performance.

In this regard, the reader is reminded to consider the *concept of independent events* [see, Chapter II: What is Gambling?, "The Odds of Winning in Gambling", supra] as an incentive to try different approaches to picking lottery numbers:

It is a common misunderstanding among many lottery players that picking the exact same numbers repeatedly for a lottery will increase the probability and/or odds for winning the lottery. Not true.

Picking or choosing the same numbers every time does not guarantee or assure that these numbers will be picked eventually.

Furthermore, if a person (for instance) picked 40 as one of their lottery numbers and one of the actually drawn numbers was 39, it does not really mean or imply that the person was <u>close</u> to winning. Even though 39 was picked and was only one number away from 40, it does not mean that after 39 was picked, 40 would be the next number picked in a subsequent lottery.

In fact, the number 40 in this scenario would have no better chance of being chosen next than any other number in the lottery pool.

In sum, this approach of conscientiousness contradicts the *cognitive flexibility* approach outlined above in (2), <u>Prime Yourself for Chance</u>, *supra*.
["Do something different", Professor Ben Fletcher, Psychologist, University of Hertfordshire, England (2010)]

(4) <u>Say Yes</u>.
Or, as now-deceased former Governor of Maryland William Donald Schaefer used to say: *Do it now!*

OK, you have followed the advice of priming yourself for good-luck ((2), above). Now, what do you do when a great opportunity comes along? The initial competing reactions of most people are *intrigue* and *anxiety*. On the one hand you are curious and even fascinated by the sudden opportunity; but on second thought you can think of lots of reasons to by-pass the opportunity.

Which impulse, intrigue or anxiety, will you act upon. Over time, each person develops his or her own <u>pattern</u>. In case you are wondering, this pattern concept explains why some people's lives seem to be full of good-luck or fortuitous circumstances, while other miserable souls are beset by doubts, failures, and regrets about choices not made and roads not taken.

Serendipitous people ((1), above) are more fearless and less regimented about trying something new, and engaging in well-considered new approaches. They think about things more than they show outwardly to others.

Having thought carefully about the matter, serendipitous people conclude, "Isn't that interesting? I think I'll try that". And they don't worry about what could go wrong.

> To be sure, the goal of life is to have good outcomes in our lives.

> Good outcomes increase our confidence.

> Good outcomes increase our self-efficacy.

> Good outcomes increase our belief that we are capable of accomplishing whatever we set out to do.

> Good outcomes also fuel our appetite for future risk.

As Professor Wiseman says: If you are truly unsure about a decision, try asking yourself, "What's the worst that can happen?" Then ask yourself, "What is the likelihood of that worst case scenario happening to me?" Further, you may want to ask, "Which of these actions will I most likely regret more in the future?"

After all, when it comes to lotteries, sweepstakes and contests, even the best of us encounter some failure.

(5) Embrace Failure.

Let's be clear. Even at the top of our game, not every lottery, sweepstakes or contest we enter will turn out well for us. Failure is a realistic option, even though we do not want it to be.

But, why do we fail? Answer: We fail so that we can evaluate, analyze and get back up to try again.

We Can Fail Even When We Think We Are Listening To The Voice From Above

---

A woman is walking along a deserted beach on a nice sunny day when she suddenly thinks she hears a deep and melodious voice from above.

"DIG!" the voice says.

Startled, the woman looks around but there is no one to be seen. I must be imagining this, she thinks. Then she hears the voice again.

"DIG!" the voice commands.

Compliantly, the woman starts digging feverishly with her bare hands, pushing away the sand. After a few minutes, and a short distance in the ground of sand, she abruptly uncovers a small elegant chest with a rusty lock.

"OPEN IT!" the voice orders her.

Well, alright, the woman thinks, I'll follow the instructions. She finds a rock, breaks the lock, opens the chest and discovers a gleaming pile of gold coins.

"TAKE THEM TO THE CASINO!" the voice commands.

With the casino not being more than a few minutes from where she was standing on the beach, the woman decided to comply. "Why not?" she thought. She goes to the nearby casino, changes all of the gold into a huge pile of roulette tokens and goes to one of the tables. All of the players, of course, gaze at her in complete disbelief. She then hears the deep voice saying to her, and to her alone:

"27. PUT IT ALL ON 27!"

The woman obeys. She bets the heavy pile of tokens on 27, as instructed from above. The roulette table groans under the weight of the tokens. Everyone is quiet in the casino. You can hear a pin drop as the nervous croupier throws the ball on the wheel. The ball goes round and round, and round, and eventually drops into 26. After the groans of the casino customers subsides, the woman hears the following from above:

"SHIT!"

[www.onlinecasinomansion.com/gambling-jokes-html]

Despite this humorous story. Embrace failure. Evaluate it, analyze it, learn from it, use *cognitive flexibility thinking,* and move forward again.

Serendipitous people are resilient.

Serendipitous people who are *cognitive flexibility thinking serendipitous people* are not always perfect and regret-free. Remember: most so-called successful people are also people who have <u>failed</u> at some point. The same is true with lotteries, sweepstakes and contest entrants.

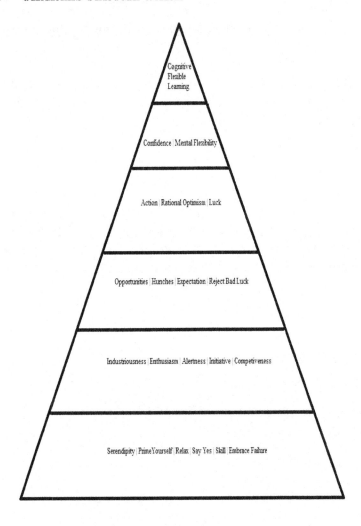

**Charles Jerome Ware's Pyramid To Success in Winning Lotteries, Sweepstakes and Contests
(January 2012)**

CHAPTER 6
HOW TO WIN LOTTERIES 102: WINNING LOTTERY NUMBER FORMULAS, PATTERNS, AND TECHNIQUES

I Have Nothing To Add

All the numbers went to a party and numbers being what they are, all the evens stayed around each other and all the odds did the same and neither group interacted with each other. Whilst two was chatting to four he noticed zero was on his own in the corner and suggested to four that because zero is sort of even he should be encouraged to mix with even numbers – four agreed. So off went two to invite zero into their little group. "Would you like to join our little group" inquired two, to which zero replied, "I have nothing to add!"

Oops! ☺

[James Glenn Davanna,
PHYJGD#NoSpam.physa.phy.hw.ac.uk]

A. Winning Lottery Number Formulas, Patterns and Techniques.

Based upon extensive research and study, and utilizing cognitive flexibility learning, I have determined that there actually are patterns, modalities, methodologies and statistical formulas for winning lottery numbers.

Patterns. Think patterns. Every lottery is a series of *patterns*.

The fact that each lottery drawing appears to be, and is touted to be, a random approximation of numbers does not mean that no *patterns* exist that cannot be used to help increase your chances of winning over a reasonable period of time.

In fact, the contrary is true. Frequent lottery players, skilled in *cognitive flexibility learning*, have discovered that when examining the same lottery game over a period of time, subtle *patterns* can be recognized that the good observer can use to his or her benefit to shift the odds of winning that lottery jackpot.

They visualize the winning numbers!

Some of the more promising tips and secrets to winning lottery number patterns, formulas and techniques are as follows:

B. Promising Tips and Secrets to Winning Lottery Number Patterns, Formulas and Techniques:

Study the history of winning combinations!

(1) Lottery patterns (or vents) do repeat themselves.
>In lottery games the patterns (or vents) that occur or happen most often are most likely to repeat themselves in the future.

This is why it is so important for serious lottery players to study past drawings and the patterns they reveal in a *cognitive flexibility learning* manner and approach. Doing this will definitely improve your chances of winning the lottery.

There is a name for this phenomenon. It is called the *frequency theory* by many. It is a well-used and tested theory that the numbers that have been the most common in the past are also going to, most likely, be the numbers that have the best chance of coming up in succeeding drawings.

(2) Do not choose all even or all odd numbers.

>*And God said, "Let there be numbers", and there were numbers.*

>*Odd numbers and even numbers created He them. And He said to them be fruitful and multiply.*

>*Then God commanded them to not monopolize against each other in the Lottery.*
>*[Charles Jerome Ware, February 18, 2012]*

Usually, all even or all odd numbers comprise only slightly more than about 5 percent of all winning lottery combinations. That is a fact.

Now then, knowing that fact, the smart or knowledgeable lottery players do not tend to choose sets of all even or all

<u>odd</u> lottery numbers. Simply put, in doing so your odds of winning the lottery are only 5% of what they would be otherwise. This is not good.

Further, mathematically and statistically, not playing all even or all odd numbers also means that your odds of winning the lottery are increased (↑)!

(3) Winning lottery number combinations are usually comprised of the entire spectrum (or universe) of possible lottery numbers, <u>from low to high numbers</u>.

That is the <u>pattern</u>. Do not forget it.

This means that you as a lottery player should <u>not</u> have all low <u>nor</u> all high numbers in your combinations.

The fact is that you have less than a 6% chance of winning the lottery with combinations of <u>all low</u> or <u>all high</u> numbers. Not good.

(4) <u>Avoid consecutive lottery number combinations</u>.

The fact is --- at least to my knowledge, and based upon my extensive research --- there has <u>never ever</u> been a winning lottery combination of 4 or 5 <u>consecutive</u> numbers. For example, no "…12, 13, 14, 15, 16…" combination of consecutive numbers.

Usually, from one to 3 numbers can sometimes work out.

By the way:

Why is the number 10 afraid of the number 7?

Answer: Because seven ate nine, and 10 is next!

[jokes4all.net/numbers.html] ☺

(5) Avoid playing previously winning lottery number combinations.

I do not understand why people do this: a lottery number wins for someone else; another player (and sometimes even the same winning player) later plays the exact same number; they lose. Wrong approach!

Past drawing history ((A), above), mathematical and/or statistical odds, and cognitive flexibility learning teach us that a snow ball probably has a better chance of making it through hell than a winning lottery combination has of repeating itself. So, govern yourself accordingly.

Note: All states publish lists of previously winning lottery number combinations on their websites, so check them for yourself.

(6) Look for trends in lottery number patterns.

Strange as it may sound, trends in lottery number combinations do develop. Look for them with a view towards anticipation of future combinations.

(7) Look for "hot numbers" and avoid "cold numbers".

The "hot numbers" and "cold numbers" theories are probably the most popular lottery number theories.

The reality is that some numbers have better success in winning lotteries than other numbers. These winning numbers are called "hot numbers". Numbers that have been least successful in winning lotteries are called "cold numbers".

The hot numbers are the numbers that are encouraged for us in the *frequency theory.*

Many lottery experts and enthusiasts agree that picking hot numbers works better than so-called random number selection.

(8) Participate in lottery pools and lottery clubs, lottery syndicates, et. al.

If you cannot find one, create or organize your own lottery pool or lottery club, or group.

The number of pools, clubs and groups of co-workers and friends who have won lotteries in recent years has been huge! Join a lottery group, or organize one!

For example, I personally knew a couple of the lottery winners in the 2010 Lansing, Michigan group that won the $129 million Michigan Lottery. This winning ticket was purchased at a Detroit-area porn shop. Remember from Chapter I, supra, the lottery does not discriminate against the location (seller) from whom you bought the winning ticket (even if the lawful seller is a "porn shop"). The lottery is about numbers only, not moral values, etc.

(9) Concentrate on lotteries that have no more than 6 numbers in them.

The reason is simple: The fewer the numbers, the better your chances of winning. There are many mega-lotteries now that have 9 or more numbers. These lotteries are naturally much tougher to win because there are more numbers to get a winning combination from.

(10) Never pick birthdays, ages, anniversary dates or other significant numbers intentionally just because you believe they are your lucky numbers.

Stick to the strategy. Always use the number patterns that you have come up with using your *cognitive flexibility learning.*

Remember (from Chapter I), numbers have no personality nor sentimentality.

By the way, if you make the mistake of <u>not</u> following my advice --- and, instead, "play your age" --- at least make sure you play your <u>correct</u> age. For example:

A woman was in a casino for the first time. The spinning ball of the roulette wheel has always caught her attention. She decides to play at the roulette table and she says, "I have no idea what number to play."

A young, good-looking man nearby flirts with her and suggests she play her age. Blushing and smiling at the man, she puts her money on number 29. The wheel is spun, and 36 comes up. The smile drifts from the woman's face and she   faints.
[www.onlinecasinomansion.com/gambling-jokes.html] ☺

Follow the tips and secrets in this book.

(11) Keep to your budget.

Serendipitous discipline! In order to win the lottery you have to play the lottery.

But, you must <u>not</u> bankrupt yourself playing the lottery, doing sweepstakes or contests, nor gambling of any type!

Keep a modest budget for these games. Concentrate on analyzing these games, and <u>not</u> just throwing cash at them on chances of winning! Do <u>not</u> go beyond your budget.

(12) Be organized. Stay organized.

For example, one simple and easy approach that worked for <u>me</u> recently was as follows:

(a) I tracked the 6-number lottery results for the previous 3 months of the lottery I wanted to enter.

(b) I picked the "hot numbers" that I <u>felt</u> had a higher probability of winning.

(c) I avoided what I determined to be "cold numbers".

(d) I chose a mixture of <u>hot</u> odd and <u>even</u> numbers across the entire field of that lottery.

(e) In the numbers I chose I picked 2 numbers that were relatively low or high in relationship and combination with the other numbers in the center of the lottery's number field.

(f) I added up the 5 numbers I had chosen. These numbers totaled between 110 and 175.

[For some reason, this appears to be a <u>winning range</u> for totaled numbers in a 6-number lottery]

(g) I won!

*(13) <u>Post-script on ((12), Be organized. Stay organized),</u>
<u>above.</u>*

I discussed my above success with a veteran lottery player and friend, and he volunteered the following information:

(a) In most lotteries that he is aware of, <u>one</u> number from the previous week's lottery is usually chosen again almost 50 percent of the time!

(b) <u>Hot</u> numbers that have not been drawn for <u>six</u> games or less account for almost 50% of the winning numbers!?!

(14) <u>Hypergeometric Distribution.</u>

This is a lottery winning technique for the mathematically-inclined lottery player.

In lottery probability theory and statistics, the *hypergeometric distribution* is a discrete probability distribution that explains the probability of K successes in a sequence of n draws from a finite population without replacement.

In other words, a random variable X follows the hypergeometric distribution if its probability mass function is given by:

$$P(X=K) = \frac{(m)(N-n)}{(K)(n-K)} \bigg/ \frac{(N)}{(n)}$$

Where,
-N is the population size (of play-a-ble numbers)
-m/N is the initial probability of success
-n is the number of lottery draws
-k is the number of successes
-(ab) is a binomial coefficient
It is positive when max $(O, n+m-N) \leq K \leq$ min (m,n) [see, "Mathematical Statistics and Data Analysis", John A. Rice, 3rd Edition, Duxbury Press (2007); "Calculation Tool for Fisher's Exact Probability Test for 2x2 Tables", K. Preacher and N. Briggs (May 2001), http://quantpay.org; "The Hypergeometric Series", W. Feller, §2.6 in An Intro to Probability Theory and Its Applications, Vol. 1, 3rd Ed. (Wiley New York); "Hypergeometric Distribution", E.W. Weisstein, MathWorld]

The Hypergeometric Formula. Let us suppose a lottery population of numbers consists of $\underline{N}$ items [let's say "50", 1 through 50], $\underline{K}$ items or numbers of which are "successes" [hot numbers]. And, for example, let's say a random sample drawn from that population of $\underline{N}$, or 50 numbers consist of $\underline{n}$ items (say, 10), $\underline{x}$ of which are "success" [hot numbers]. Then the *hypergeometric probability* for a winning lottery number is:

$$\textbf{H(x; N, n, K) = [kCx]/[NCn]}$$

As you can see, the hypergeometric formula is not easy to maneuver, and it requires many hours of time-consuming computations: as well as mathematical and statistical ability. **BUT IT WORKS!**

EXAMPLE Use Of The Hypergeometric Distribution
Consider, for example, that 5 lottery numbers are chosen randomly from the total of 10 lottery numbers without repetition. We now want to calculate the probability of get-

ting exactly 2 winning numbers out of the total of 6 winning numbers from the universe total of 10 numbers.
The Hypergeometric Distribution Formula is:

$$h(x;N;n;k) = [_kC_x] [_{N-k}C_{n-x}] / [_NC_n]$$
where,
N is the total population size,
n is the total sample size,
k is the number of selected items from the population size, and
x is a random variable.

First Step in calculating the probability:

Find $[_kC_x]$

where, ◆ N=10, n=6, k=5 and x=2

$$[_kC_x] = (k! / (k-x)!) / x!$$
$$= (5! / (5-2)!) / 2! = 20 / 2 = 10.$$

Second Step:

Find $[_{N-k}C_{n-x}]$

where, ◆N-k=5 and n-x=4

$$[_{N-k}C_{n-x}] = ((N-k)! / ((N-k)-(n-x))!) / (n-x)!$$
$$= ((5! / 1!) / 4!) = 5 / 4! = 5.$$

Third Step:

Find $[_NC_n]$

where, ◆N=10 and n=6

$$[_NC_n] = (N! / (N-n)!) / n!)$$
$$= ((10! / 4!) / 6!) = 151200 / 6! = 210.$$

Fourth Step:

Find $[_kC_x] [_{N-k}C_{n-x}] / [_NC_n]$

where,

$[_kC_x] = 10$, $[_{N-k}C_{n-x}] = 5$ and $[_NC_n] = 210$.

$$h(x;N;n;k) = [_kC_x] [_{N-k}C_{n-x}] / [_NC_n]$$
$$= [_5C_2] [_5C_4] / [_{10}C_6]$$
$$= (10 \times 5) / 210$$
$$= 0.238.$$

Hence there is a 23.8% possibility of choosing exactly 2 winning numbers without repetition.

Dr. Joan Ginther: "The Luckiest Woman In The World"

Ask those knowledgeable in Texas and California about Professor Joan Ginther, who earned her Doctorate degree (Ph. D.) in mathematics and statistics from California's Stanford University and won the Texas Lottery a whopping 4 times --- winning a total of $21 million!
Understandably, Dr. Ginther declines to be interviewed about her wins. But other mathematical experts say that the chances of Dr. Ginther accomplishing such a feat by sheer luck or chance are no less than <u>one</u> in 18 septillion! That's <u>one</u> in 18,000,000,000,000,000,000,000,000!!!
[<u>see</u>, www.dailymail.co.uk/August9th,2011; www.huffingtonpost.com/July13,2010; beaumontenterprise.com, August 9, 2011, **"God or Math: Texas woman who won lottery four times is a Math GENIUS!"**]

Frequently called the *"luckiest woman in the world"*, Joan R. Ginther won her four fortunes worth millions of dollars over a span of fifteen years. Three of her wins were purchased at the same mini-mart in the same small town of Bishop, Texas. Her home, though, is in Las Vegas, Nevada.

The residents of little Bishop, Texas seem to believe that God has been behind Dr. Ginther's unbelievable luck. Some experts, however, prefer to think that more than luck is behind Ginther's remarkable lottery success. They think: perhaps this Ph. D. mathematician and statistician has cleverly unlocked the mathematical *algorithms* that determine what and where winning numbers are in each game.

The Texas Lottery Commission has declared that Ginther must have been "born under a lucky star". But, Professor Nathanial Rich of the Institute for the Study of Gambling & Commercial Gaming at the University of Nevada in Reno is not so sure.

As he reported in the August 2011 issue of *Harper's* magazine:

The odds of Ginther accomplishing this feat (having this phenomenal luck) are one in 18 septillion: that's <u>one</u> in 18,000,000,000,000,000,000,000,000!!!

Putting this into perspective, according to Professor Rich:
- There are only one septillion stars in the universe!
- There are only one septillion grains of sand on Planet Earth!
- And, Ginther's luck can only happen <u>once</u> in 1,000,000,000,000,000 years!

Using a personal example as an illustrative comparison, Professor Rich opines:

- "A few months ago, I found a $100 bill lying on the ground in JFK terminal (in New York). The odds of that happening is something like one in 100,000. But this is ridiculous."
["Meet The 'Luckiest' Woman In The World", Kiri Blakeley, 07/21/2011, pro-files.google.com/102871322541498553806/buzz; "Professional Statistician from Texas Wins Lottery Four Times", 08/9/2011, www.straitstimes.com/BreakingNews/ANN/Story]

Dr. Joan R. Ginther's Doctoral Dissertation

By the way, for those curious and serious lottericians, Joan Ginther's doctoral dissertation for her Ph. D. was not on the subject of the lottery.

In researching information and background for this book, my legal assistant and I actually found and read the dissertation, titled "Pretraining Chicano Students Before Administration Of A Mathematics Predictor Test" (A major portion of the dissertation, Evergreen Valley College and Stanford University, 1976).

Ginther's thesis does, however, support my contentions regarding the importance of cognitive flexibility learning in winning lotteries, sweepstakes, and contests.

The Ginther Algorithm

What is an **algorithm (al-go-rithm).**

An <u>algorithm</u> is defined as a step-by-step problem-solving procedure; especially an established, recursive computational procedure for solving a problem in a finite number of steps. Like a lottery.

Mathematical algorithms are of wide application in science and engineering. Basic algorithms for mathematical computation include those for generating random numbers, performing operations on matrices, solving simultaneous equations, and numerical integration.

It appears to me, from my study, that Dr. Ginther may have used a matrix of factors in her lottery analysis algorithms, including:

(1) "Hot" and "cold" number analysis; (supra, B(7); B(15), infra)
(2) John Nash's game theory; (infra, Nash)
(3) Distribution of winning lottery numbers over time; (infra, (19)(b))
(4) Winning number frequency (similar in several respects to "hot" and "cold" number analysis); (infra, (19)(f))
(5) Adjacent pairs pattern analysis;
(6) Elapse time analysis (Data Gathering);
(7) Repetition pattern analysis; and
(8) Cognitive flexible learning. (supra, Chapter V)
    And, I believe, Dr. Ginther utilized computers to create her lottery number algorithms.

(15) Create a "Hot Numbers" or "Hits" Sheet or Chart.

The *"Hot-number" theory* and the *"cold-number theory"* are probably the most popular lottery theories.

This sheet reflecting the frequency of "hot" numbers or "hits" of prior winning numbers in your lottery of choice can be a chart, a graph, or whatever. But, it should be *VISUAL!* The Sheet or Chart should make it easy to see and visualize the winning history of the lottery numbers you are considering. For example, take this oversimplified chart:

NUMBERS "HOT" NUMBERS or "HIT" NUMBERS CHART

NUMBERS	1	2	3	4	5	6	7	8	9	10	11
10				X				X			
12	X		X								X
18					X				X		
25		X	X							X	
27					X						
32		X			X	X	X	X			
33	X									X	X
39				X					X		
40					X			X			
43		X	X	X					X	X	
49	X					X		X	X		

Week of Lottery

The above **"Hot" numbers or "Hit" numbers chart** in the proposed 6-number combination lottery reveals, inter alia, the frequent "hot" or "hit" numbers to be:

12-25-32-33-43-49

(16) The Number Difference Formula.

For those lottery novices who are struggling to develop or find their own formula for picking winning lottery numbers, you may want to start out with a very simple and easy approach such as, what I call, the *number difference formula.*

There are several variations of this approach being used, but the fundamental essence of the approach is as follows (using, for example, a winning Maryland Lottery number):

a. The winning lottery number is: 04-10-16-38-48-34 (indeed, a rare combination of all even numbers)

b. I now calculate the <u>differences</u> between each lottery number and the previous number, and I get: 06-06-22-10-14 (another rare combination of all even numbers)

c. The psychological value of this approach for novice lottery players is that the player has the sense that he is playing off of proven winning lottery numbers, and thus feels better about the game, etc.

(17) Reputable Software Programs are OK if you cannot, on your own, determine your lottery numbers. However, do <u>not</u> spend a lot of money on these programs. They probably will <u>not</u> make <u>that</u> much of a difference in your chances of winning.

It is your personal research, hunches, pattern-finding, and so forth, that help the most.

(18) Rational Persistence.

Persistence counts in winning lotteries, sweepstakes and contests.

The chances of winning (under certain conditions) are persistent; the "hot" numbers are persistent; the "cold" numbers are persistent. You must be rationally persistent as well.

(19) Pattern Analysis.

Pattern Analysis is another name for a player's efforts in identifying trends and patterns of winning lottery numbers. It utilizes most, if not all, of the tips and secrets I have presented and discussed before in this book.

In a nutshell, *pattern analysis* consists of the following:
- distribution of winning numbers over a period of time,
- behavioral stratification of numbers based upon numerical position,
- relationship between mutually winning numbers, and
- winning number frequency.

[www.omninerd.com/articles/PatternAnalysis]
Steps in winning pattern analysis include:

a. Data Gathering. Gather the data and history on the winning lottery numbers. I recommend you go back at least 12 months in collecting the winning number combinations.

b. Distribution of Winning Lottery Numbers Over Time.

Check to see if there is a trend for the winning number combinations to occur with an even distribution (particularly with MegaMillion, PowerBall, and other ball lottery games). Otherwise, simply keep track of the frequency of winning numbers as demonstrated previously in this chapter.

c. Behavior Stratification of Lottery Numbers Based Upon Numerical Position.
   At a more indepth level, the more advanced lottery player looks positionally at the location of winning numbers from past drawings. For example, during one 2-month stretch of a lottery recently, I noticed that the following numbers came up in this exact same order in 4 out of 10 lottery drawings:
   "42-44"
   This was a pattern.

d. Relationship Between Mutually Winning Numbers.
   Analyze which numbers tend to win together. For example, in a trend of lottery combinations, how many winning 5-number combinations have the numbers "35, 42 and 44" in them.

e. Common Differences Between Winning Numbers.
   What are the odds or chances for "even" versus "odd" numbers being in the winning combination. Study of past games will reveal this answer.

f. Winning Number Frequency.
   Simply put, what is the temporal frequency by which a numbers wins. Did the winning number occur multiple times within the past 12 months, or did it come up multiple times 5 years ago? There is a difference. More recent frequency of winning numbers is more compelling.

[see, The Frequency Theory]
[www.luck-lotto-numbers.com/lotterynumberpatterns; usa-dailynews.info/thesecrettheoriesbehindwinningthelottery; ezinearti-cles.com/LearntheTheoriesinWinningtheBigLotteryGame; www.ehow.com/how-winninglottery, 4805723; www.comparelotto.com/usingtheprobabilitytheoryinlottery. php; www.lotterynumberadvisor.com/lotterynumberpatterns; www.use4.com/lotto.html; www.smartluck.com/3-winninglottomethods.htm; www.lottery.com/lottery-winningnumbers.cfm; www.lotterynumberadvisor.com/lotto; Advanced Lottery Theory (A Brief Summary), Jim Schwartz, Satori Publishing (2008); "The Lottery is Rigged", Uncoverer (Accessed October 2007), www.uncoveror.com/lottery.htm; "How to Play: Play the Game", MegaMillions.com (Accessed October 2007)]

(20) Neural Networking, or "Parallel Processing".

As we have discovered, winning the lottery involves finding as many subtle non-random influences as possible on the numbers. *Neural Networking, also known as Parallel Processing*, is another technique used for that purpose.

Neural Networking (or Parallel Processing) is said to be a type of artificial intelligence that attempts to imitate the way a human brain works. Its primary purpose is to find hidden "patterns of chaos". Neural networking is particularly effective in *predicting events* when there is a large database of prior examples (prior data) to draw upon. In other words, for our purposes here, using neural networking to analyze a large database of winning lottery numbers can help you win the lottery!

John Forbes Nash, Jr.

The best example of the human brain exemplifying neural networking, that I am aware of, is the great winner of the prestigious Nobel Prize of 1994 in Economics, John Forbes Nash, Jr.

Dr. Nash, the king of "Game Theory" and a brilliant mathematician, had the uncanny ability to visually (and mentally) see patterns in encrypted messages that were, for all practical purposes, "invisible" to the average person. Effectively, Nash was a human computer. He is considered by many experts to be one of the greatest mathematical minds of the 20th Century.

John Nash visualized number patterns!

His tremendous mental exploits were featured in the best-selling 1998 book, "A Beautiful Mind", by Sylvia Nasar, as well as the very popular Academy-Award winning movie of the same title, and starring actor Russell Crowe as Nash.

Study the history, trends, and patterns of prior lottery drawings and, using neural networking, *visualize* the hidden or weighted influences that take the randomness out of the game.

[A Beautiful Mind (The Story of John Forbes Nash, Jr.), Sylvia Nasar, Simon & Schuster (1998); "A Celebration of John F. Nash, Jr.", a special volume, Duke Mathematical Journal, Vol. 81, No. 1, Duke University Press (1995); "Parallel Control", John Nash, RAND Memorandum, No. 1361, 8.7.54; Plenary Lecture, Madrid, Spain, 8.26.96, op cit; Advanced Lottery Theory (A Brief Summary), Jim Schwartz, Satori Publishing (2008)]

What is John Nash's "Game Theory" and How does it help you win Lotteries and Sweepstakes?

Game Theory is a mathematical method or approach for analyzing calculated circumstances such as in games (including lotteries, chess, etc.), where a person's success is based upon the choices of others. It essentially removes the concept of randomness.

Game Theory is principally used in economics, political science, and psychology, and in other, more prescribed sciences such as logic and biology. More technically, it is described as the study of mathematical models of conflict and cooperation between intelligent rational decision-makers.

Another name for Game Theory is *Interactive Decision Theory*.

From my perspective, what game theory does is to unify and systematize your *cognitive flexibility learning*-developed intuition into winning strategies.

Creating, developing and using matrixes, charts, and graphs in game theory or interactive decision-making is important, as we have discussed in this book.

Follow the tips and secrets in this chapter and win. Be patient.

A normal, or strategic, form game is usually represented by a matrix which shows the players, strategies, methodologies, and payoffs, as follows below. In the matrix below, assume there are only 2 players. One chooses the rows and the second player chooses the columns. The payoffs are provided in the interior of the matrix. The first number is the payoff received by the row player (Player 1), and the second number is the payoff for the column player (Player 2). Assume that Player 1 plays *Up* and the Player 2 plays *Left*.

In that case, Player 1 gets a payoff of 4, and Player 2 gets a payoff of 3. The assumption is that each players acts simultaneously (at the same time) with no advance knowledge of each other's choice.

['Game Theory: Analysis of Conflict", Roger B. Myerson, Harvard University Press (1991); "Game Theory", Introduction, The New Palgrave Dictionary of Economics, 2nd Edition (2008; "Game Theory", Drew Fudenberg & Jean Tirole, MIT Press (1991); "Essentials of Game Theory: A Concise Multidisciplinary Introduction", Kevin Leyton-Brown & Yoav Shoham, Morgan & Claypool Publishers: San Rafael, CA (2008); Games of Strategy, Avinash Dixit and Susan Skeath, Norton Pub.: New York (1999)]

(21) Using Lottery Groups or Syndicates to Win

Playing the lottery, and more importantly winning the lottery, is all about the **numbers**. To that end, it simply makes sense that the more people working together to play, and win, the lottery is important. Consequently, lottery players should, when conveniently and reasonably possible, **pool** their knowledge and financial resources with other similar-minded players and form **lottery groups** or so-called **lottery syndicates**. For example, take the case of the "Lucky 12".

(a) *The "Lucky 12"*

On Friday, January 4th, 2008, a group of twelve teachers and secretaries from Denville, New Jersey's Morris Knolls

High School took a fateful bus ride to the state capitol city of Trenton in the early afternoon. Calling themselves the "Lucky 12", the group turned in a winning lottery ticket worth more than $49.5 million before the necessary twenty-five percent withdrawal in federal taxes.

The "Lucky 12" had won half of the December 18th, 2007 New Jersey Mega Millions drawing. They had purchased the winning ticket at a convenience store and had decided to choose the cash option.

Using several related and different methodologies (all of which are discussed in this book), each member of the "Lucky 12" had chosen and submitted on behalf of the group a different number. To my knowledge, no one had chosen random selection as an option.

Upon my best information and belief, and back-up research, the methodologies used by members of the winning group were:
- Cognitive Flexible Learning (not surprising for *teachers*);
- "Hot" and "Cold" Number theories;
- Game Theory (Interactive Decision Theory);
- Winning Number Frequency; and
- Pattern Analysis;

Apparently, the "Lucky 12"lottery group had been pooling their money and playing the New Jersey Mega Millions lottery "as a tradition" for at least 13 years. Among the winners in the group were several veteran teachers who were close to retirement age.

(b) *The "DPD 18"*

Interestingly, in another demonstration of lottery group power, 18 members of the Denville, New Jersey Police Department bought the winning lottery ticket for a $27.2 million New Jersey lotto jackpot in 1994.

Upon information and belief, the police department personnel used principally just three (3) methodologies:

- "Hot" and "Cold Number theories;
- Trends in lottery number patterns; and
- Distribution of Winning Lottery Numbers over Time.

But, they won!

(c) *"The Martinez-Sanchez Lottery Win"*

On March 29th, 2008, there was only one winner in the $13 million Lotto Texas jackpot. It was a company: Martinez-Sanchez & Associates, Ltd., of McAllen, Texas.

The lotto ticket was purchased in Pharr, Texas by the company with what is called a Cash Value Option ("CVO").

The lucky numbers drawn were **11, 22, 40, 27, 33, 2.**

The winning check was collected by the company's general partner, Martinez-Sanchez Management, LLC, by way of its manager M. Martinez Sanchez. It amounted to a one-time lump payment of $8,644,766.00, before taxes. The winner requested "minimum publicity."

These winners are reported to be computer and numbers savvy. And lucky.

(d) The Famous, or Infamous, Virginia Lottery Syndicate of Melbourne, Australia

Last, but certainly not least, in this issue of using lottery groups or syndicates to win is the concept of buying <u>all</u>, <u>most</u>, or <u>many</u> of the number combinations in the pool to win. In other words, a "buy-out" of the lottery.

One example of this happening was the 1992 win by a Melbourne, Australia-based syndicate of 2,500 investors who won the Virginia Lottery jackpot of $27 million. Even though the group (International Lotto Fund) fell short of purchasing all 7 million possible ticket combinations, they bought enough of them to win.

Additional particulars in the Australian syndicate's win are as follows:

- The group's $5 million purchase of the $1 tickets cornered 5 million of the 7 million number combinations for the February 15, 1992 Virginia Lottery.

- Apparently, only a shortage of purchase time prevented the Australian investors from buying the remaining 2 million combinations.

- All possible combinations of the 6-numbered lottery were derived from numbers 1 to 44.
- The winning lottery numbers were: **8, 11, 13, 15, 19 and 20.**
- The lucky ticket was purchased at a Farm Fresh convenience store.

It was a big risk for the 2500 investors, but their $5 million investment paid-off with a $27 million reward.

<u>Caveat: Be Careful! Do not do this alone! Use a group or syndicate! Going solo in such an undertaking can be fatal to your financial health!</u>

Take the case, for example, of the retiree in 1990 who walked into a hardware store in Sacramento, California, with a diaper bag full of $20 bills to spend on the California Lottery. The store's employees spent all night printing out the woman's 30,000 lottery tickets for what was then a record California lottery jackpot of $69 million! She did not win.

Further, with the unfortunate notoriety she generated in having lost over $30,000 of her hard-earned retirement funds, she has not been heard from since.

(22) Winning "Pick 3" and "Pick 4" Lottery Games

For fun and entertainment, try the following system I used and won last month:

a. Create the Chart (the middle portion) (B), infra.

b. Post (for example) the last 6 winning Pick-3 numbers above (A). The more games you pick the better.

c. Check-off the numbers in B as they are reflected from the "winning numbers" in A.

d. Add the frequency of the numbers and post the highest frequency of numbers to the right (C) as your current winning Pick-3 numbers.

Winning Pick-3 and Pick-4 Lottery Games

A	(505)	(849)	(652)	(543)	(419)	(438) **Winning**
	#1	#2	#3	#4	#5	#6 **Pick-3**
	↓	↓	↓	↓	↓	↓ **Numbers** ↓

(B)	0✓	0	0	0	0	0
	1	1	1	1	1✓	1
	2	2	2✓	2	2	2
	3	3	3	3✓	3	3✓
	4	4✓	4	4✓	4✓	4✓ } (C) **549**
	5✓✓	5	5✓	5✓	5	5
	6	6	6✓	6	6	6
	7	7	7	7	7✓	7
	8	8✓	8	8	8	8
	9	9✓	9	9	9✓	9✓

	↑	↑	↑	↑	↑
	#1	#2	#3	#4	#5

PAST GAMES

↑ *CHART FOR WINNING PICK-3 AND PICK-4 LOTTERY GAMES* ↑

C. How to Form a Lottery Club, Pool, Group or Syndicate.

A lottery club, pool, group or syndicate is formed by at least 2 or more people whose purpose is to pool or combine their money to purchase lottery tickets with the intent and agreement to split or share the winnings.

[see, (21) Using Lottery Groups or Syndicates to Win, *supra*]

These clubs or pools can have as many members as they wish, and they can play as many or as few lotteries as they want. It is crucial, though, that the group decide in advance how it will distribute the winnings among the members.

Some additional basic advice for forming a successful lottery club, pool, group or syndicate is as follows:

1. Organize the group with like-minded people (people you get along with) such as family, friends, co-workers, etc.
2. Determine and set the rules for your group early in the process.
3. Make sure all agreements and understandings among the group are in writing and signed by the members. These writings should be treated as binding contracts. This is very important.
4. Your club or group must acquire an Employer Identification Number (EIN) to be recognized as legal.
5. Play the lottery. You must play to win.
6. Follow the other instructions, advice and tips in this book.

CHAPTER 7
HOW TO CREATE YOUR OWN WINNING LOTTERY ALGORITHM

A. *Algorithm*
(Definition) For our purposes in this book, an "algorithm" is a logical and rational step-by-step procedure or process for picking winning lottery numbers.

In this chapter we will tie together the methodologies we have learned in previous chapters to manually (without a computer) fashion together winning lottery numbers. We will create an <u>example</u> for teaching purposes only.

B. *Example for Creating a Lottery Number Algorithm*

1. Choose a lottery --- say, "The Fran Lottery", for example.
2. Research the winning lottery numbers in "The Fran Lottery" over a period of time --- say, the past 3 months, for example.
3. Compile or list the winning lottery numbers in "The Fran Lottery" for the past 3 months, in the <u>order</u> (chronological order) they appeared and won:

Draw Dates	Winning Lottery Numbers
2/21/2012	9-30-39-42-47-37
2/18/2012	23-28-50-56-59-5
2/15/2012	11-12-32-52-56-11
2/11/2012	1-10-37-52-57-11
2/8/2012	20-22-32-41-52-40
2/5/2012	9-30-39-42-47-37
2/1/2012	1-19-32-41-42-11
1/28/2012	2-23-24-37-38-18
1/25/2012	16-25-28-32-40-3
1/21/2012	3-5-10-26-27-27
1/19/2012	3-4-18-29-50-20
1/16/2012	17-23-30-37-45-4
1/12/2012	11-24-26-36-38-13
1/7/2012	8-12-25-33-45-5
1/4/2012	3-5-16-25-37-16
12/31/2011	1-2-18-34-43-27
12/28/2011	9-14-16-25-33-55
12/24/2011	8-23-29-32-37-31
12/21/2011	14-18-20-24-34-42
12/17/2011	2-32-33-35-38-32
12/14/2011	15-23-43-45-56-7
12/10/2011	17-28-38-39-51-33
12/7/2011	8-13-71-34-59-22
12/3/2011	5-33-41-54-59-13

4. Look for lottery number patterns (or vents) that re-peat themselves: none identified in our example.
5. Look for trends in the lottery number patterns:For instance in these 24 drawings over a 3-month period:

16/24 (67%) of the first-drawn numbers are odd numbers.
14/24 (58%) of the 2nd-drawn numbers are even numbers.
15/24 (63%) of the 3rd-drawn numbers are even numbers.
13/24 (54%) of the 4th-drawn numbers are even numbers.
14/24 (58%) of the 5th-drawn numbers are odd numbers.

16/24 (67%) of the 6th-drawn numbers are <u>odd</u> numbers.

Therefore, we see that the <u>first</u> and <u>last</u>(6th) numbers of the lottery are dominated (67%) with <u>odd</u> numbers, and we see that the middle lottery combinations (2nd, 3rd, 4th numbers) are dominated with <u>even</u> numbers.

There are other trends in the lottery combinations in this example but, for the sake of brevity, we will not discuss them as part of this instruction.

    6. Look for the "hot" numbers and the "cold" numbers. <u>Play</u> the "hot" numbers. <u>Avoid</u> the "cold" numbers [Chapter VI, Section B, (7) "Look for 'hot numbers' and avoid 'cold numbers'", <u>supra</u>]:
        (a) The "hot" numbers are the numbers in the winning lottery combinations that are played most often. They are as follows:
        - The <u>top</u> 6 "hot numbers in our example are:
          5-11-23-32-33-37
        - The <u>next</u> 4 "hot" numbers in our example are:
          16-18-25-38
        (b) The "cold" numbers in our example are, among others: 6, 7, 15, 19, 21, 31, 43, 44, 48, 49, 51, 54, 55, 57, 58, 60, etc.

CONCLUSION

The winning lottery combination we have derived with our <u>manual</u> (no use of computer) <u>algorithm</u> in this example is:

5-11-23-32-33-37

Adding these six lottery numbers up, we get a total of <u>141</u>, which is well within our *winning range* of "between 110 and 175" [Chapter VI, Section B, (12) "Be organized. Stay organized", <u>supra</u>].

Congratulations!!! You have won!

C. *Example for Creating "Pick-3" (and even "Pick-4") Number Lotto Algorithms Using Logic*

<u>Logic</u>

Logic is the formal systematic study of the principles of valid inference and correct reasoning.

Logic is the study of necessary truths and of systematic methods of clearly expressing and rigorously demonstrating such truths.

["Logic Matters", Peter Thomas Geach, University of California Press (1980); "Introduction to Elementary Mathematical Logic", Abram A. Stolyar, Dover Publications (1983); "Is Logic Empirical?", H. Putnam, <u>Boston Studies in the Philosophy of Science</u> (1969); "Logic and Ontology", T. Hofweber, <u>Stanford Encyclopedia of Philosophy</u> (2004), Edward N. Zalta, Editor]

Probably one of the most elementary and often-used examples of logic is the following:

- - All babies are cute.
- - I have a baby.
- - Therefore, my baby is cute.

Whether or not the above example (a <u>syllogism</u>) is true or false, it is still logical as presented; even though many would conclude it is false.

A syllogism, such as the example above, is composed of 3 statements:

(1) The major premise, or general observation (which does not necessarily have to be true)

(2) The minor premise, or particular observation (which is generally true); and

(3) The conclusion, which is something that someone might correctly deduce from the (1) major premise and (2) minor premise stated.

(4)

Let's try another example or syllogism:

- Everybody who has been exposed to the Z-factor virus has died from it.
- Newt Q. has been exposed to the Z-factor virus.
- Newt Q. will die because of the Z-factor virus.

What is the difference between the two syllogisms? It's very clear that in the first syllogism, the major premise is probably not true. Surely there are babies in world who are not cute. On the other hand, the major premise of the second syllogism we can probably accept as true. While there may in fact be people who have been exposed to Z-factor and lived, we have no record of them. On the other hand, every Z-factor case that we've seen has resulted in death. Therefore, we can proceed fairly confidently from our major premise to a conclusion that is sound.

A Basic "Pick-3" Number Example Using Logic

You are playing the "Pick 3" Number Lotto using four prior winning number combinations as history for the drawing:

1. The available numbers to choose from are 1 to 9 (1, 2, 3, 4, 5, 6, 7, 8 and 9).
2. The winning three numbers for the <u>first</u> drawing are "137".
3. The winning number combination for the <u>second</u> drawing is "347".
4. The winning three numbers for the <u>third</u> drawing are "468".
5. The winning combination for the <u>fourth</u> drawing is "137".
6. Assume that two of the winning combinations have the same total sum value (when the 3 numbers are added up).
7. Also, assume that all of the odd numbers in the game (1, 3, 5, 7 and 9) are drawn, including the current drawing.
8. *What is a likely winning drawing combination in the current drawing you are playing?*

CONCLUSION

The likely winning "Pick-3" Number Lotto combination in this example is:

"459"

(a) The total sum value of these winning number is 18; the same as the total sum value of the winning "Pick-3" numbers in the <u>third</u> drawing.

(b) The 9 is the only odd number not drawn in the previous four drawings, which in combination with 4 and 5 totals 18.

Congratulations!!! You have won this "Pick-3" Lotto drawing!

CHAPTER 8
HOW TO WIN SWEEPSTAKES 101

Sweepstakes Practices

The U.S. Senate is investigating deceptive sweepstakes practices by certain organizations.

These organizations target the elderly and make them think they will receive a lot of money, but in reality they never see any of it.

The most popular of these scams is called Social Security.

[www.agehumor.com; getamused.com] ☺

Cute and funny. But Social Security is <u>not</u> a sweepstakes. And sweepstakes are generally not scams. I am hopeful that Social Security will not be a scam when I become of age to collect it.

I. The Psychology of Winning

Most winners of lotteries, sweepstakes, and contests have what I call a psychology of winning.

For example, in the many instances when I have won I have always had *"COPE"*:
 <u>C</u>onfidence of my success!
 <u>O</u>rganization for the game!
 <u>P</u>atience for the game!
 <u>E</u>njoyment for the process of winning!

 I have always <u>avoided</u> the following when enjoying and playing lotteries, sweepstakes and contests:

 <u>G</u>reed.
 <u>E</u>ntitlement (feelings of entitlement).
 <u>E</u>nvy of others who win.

If you do not <u>enjoy</u> playing the sweepstakes, lotteries, and contests --- <u>do not participate</u>!

Lotteries, sweepstakes and contests are mind games. Your beliefs and your attitude are very important if you want to win consistently. You must believe you will win in order for you to win.

II. Your Odds of Winning a Sweepstakes

Like any other hobby or field of interest, there are different views and opinions among sweepstakers concerning the frequency of entering to win.

The individual opinions on this "frequency of entering" issue can be diverse and strong but statistics tend toward the middle range.

QUESTION

For example, in a random sweepstakes drawing with 50 entries, what would you say the odds are of winning under the following conditions?

    A. If you enter 1 sweepstakes 3 times;

    B. If you enter 3 sweepstakes 1 time.

ANSWER

In <u>A</u> above, you have three entries out of 50, so your chance of winning assuming all drawings have equal chance of being picked is:

$$3/50 = 0.06, \text{ or } 6\%.$$

In B above, if you enter 3 sweepstakes 1 time, and you assume each sweepstakes is the same, then your chance of winning at least one of them is the complement to losing all of them. This probability is given by:

$$1 - (49/50)^3 = 7351/125000 \approx 0.058808, \text{ or } 5.8\%.$$

Therefore, it appears you have a slightly better chance of winning at least <u>once</u> if you put all three of your bets in one sweepstakes. But, as you see, the difference is small or marginal.

[http://math.stackexchange.com/questions/24281/odds-of-winning]

III. *Methodologies for Winning Sweepstakes*

A. The 10-Step Plan.

(1) Choose a sweepstakes to enter. There are numerous ways to find sweepstakes to enter:
- The Internet.
- Sweepstakes publications (listing many different kinds of sweepstakes, contests, et al).
- -Magazines and other publications.
- -Stores, shops, other retail establishments.
- -Television commercials, etc.
- -Radio programs, etc.

(2) Choose the more difficult or complex sweepstakes.

As you gain more experience in sweepstaking, choose some more difficult sweepstakes to enter. Why? Because the more difficult sweepstakes will invite fewer entries and participants. More difficult sweepstakes are like "money in the bank" for experienced sweepstakers. You will have far better odds of winning because of a shortage of competition.

(3) Think a little bit when choosing which sweepstakes to enter.

Not all sweepstakes are the same. When you become more experienced you will inherently begin to see the differences and choose the right ones for you. For example:

- I, as a rule, never enter a so-called sweepstakes that requires I buy something in order to enter.
- I, as a rule, never enter an alleged sweepstakes that requires I complete some type of program (which, again, usually means spending more money).

- I, as a rule, <u>never</u> enter a sweepstakes that "pops-up" on my computer; regardless of whether it is a free laptop, ipod, television, or even a car.

<u>Remember:</u> There are plenty of good, free sweepstakes in the marketplace that do not require you to buy something or jump through hoops.

(4) <u>Familiarize yourself with rules of the sweepstakes and follow them carefully.</u>

In order to win the game you have to play the game; in order to play the game you have to follow the rules of the game. If you do not follow the rules precisely or exactly, your entries will be promptly disqualified.

<u>TIP</u>: Now, for the good news. Since many people <u>fail</u> to follow the rules exactly, this fact increases (↑) your chances of winning!

In sweepstakes, the rule-followers win!

(5) <u>Follow the entry deadline.</u> Almost universally, all sweepstakes have an entry deadline. Sweepstakes winners always follow the rules and <u>beat the deadline</u>!

(6) <u>Print legibly.</u> Make sure you print your information legibly on your entries. And, otherwise, make sure your writing is easily readable. Do not create an excuse of illegible handwriting (or handprinting) for not winning the sweepstakes.

If your handwriting or handprinting is awful or atrocious, consider typing your information on entries (if the rules permit it).

(7) <u>If an answer is required by the sweepstakes, always</u> <u>provide the RIGHT ANSWER.</u>

If the sweepstakes asks for an answer and you do not know the <u>correct</u> answer, do not waste your time entering that sweepstakes. What's the point?

(8) <u>Enter as many times as the sweepstakes allows (and as</u> <u>you can conveniently and affordably do).</u>

The general rule is that the more times you enter a sweepstakes, the better your chances of winning. But, do not get carried away. Be sensible.

<u>Never</u>, however, enter more times than the rules allow. You <u>will</u> be disqualified if you do so.

(9) <u>For paper mail-in entries, distinguish them in some</u> <u>way.</u> And, always use the correct <u>paper</u> or <u>card</u> size for your entries.

For example, use <u>color</u>, <u>envelope size</u>, other <u>designs</u> on envelopes, etc., to distinguish your entries from the pack of other entrants. Use larger envelopes (if the rules permit); decorate your entries; and/or fold your entries in a unique manner.

(10) <u>Once you have won, follow the requirements to claim</u> <u>your prize.</u>

This process always requires (i) responding to the winning notice in time (another deadline), and (ii) it usually requires proper execution of an affidavit. Frequently, your signature on the affidavit must be notarized.

TIP: However, beware of the many scams and schemes that falsely say you have won something, but they just want your money or your identity.

Remember: If you are asked to pay money to claim your prize, forget it. It's a scam.

Another tip for Online Sweepstakes. If you want to enter online sweepstakes, seriously consider getting a separate email account just for that purpose. Do not mix your online sweepstakes activities with your business email or personal email.

[www.wikihow.com/win-sweepstakes; "An Introduction to Sweepstakes and Contests Law", Steven C. Bennett, Esquire, The Practical Lawyer (August 2007); Also see, *Seattle Times Co. v. Tielsch,* 495 P.2d 1366 (Wash. 1972): the "free" entry form must not be unduly burdensome]

<h2 style="text-align:center">B. The 25-Step Plan.</h2>

Companies that offer sweepstakes and contests actually WANT you to win. Why? Because they are joyfully, voluntarily and enthusiastically promoting themselves, their products and/or services.

And, people like you and me win sweepstakes everyday. Why? Because we try. We put forth the effort to win.

You can win sweepstakes consistently if you do the following:

   (1) Remember the "4 P's" for winning sweepstakes:
     • Patience: Do not expect to win quickly.
     • Perseverance: If you enjoy it, stick with it.
     • Positive Attitude: Important in everything we do.

- Postage (less so now with the Internet).

(2) <u>Goals</u>. Decide what your realistic goals are for entering sweepstakes: money, a car, a house, a hobby, friends, a vacation, fun, etc. Why are you doing this? Set goals.

(3) <u>Pick the sweepstakes you want to enter and learn the rules.</u>

(4) <u>Follow the rules carefully for each sweepstakes.</u> Rules... Rules... Rules. Follow the yellow-brick rules!

(5) <u>Set up a separate sweepstakes email address.</u>

This way you can better recognize wins, avoid spam emails, detect scams, etc.

(6) <u>Consider downloading Roboform and Texter programs</u>, particularly for the Internet sweepstakes.

(7) <u>Set aside special time to enter sweepstakes on a daily or weekly basis.</u> Set goals for the number of entries per sweepstakes, strategize, etc.

(8) <u>Enter each sweepstakes quickly upon learning of the time frame for entering.</u> Do not miss the deadlines.

(9) <u>Depending upon your goals and the prizes,</u> do not overly restrict yourself to just "daily", or "one entry only", or "multiple entries" sweepstakes. Consider entering them all!

(10)   <u>Strategize each sweepstakes.</u> Think about each one --- at least a little thought!

(11)   <u>Avoid disqualification.</u> At all costs. Follow the rules precisely, and do not get disqualified. If you, for some reason, cannot follow the rules --- do not waste your valuable time entering that particular sweepstakes.

(12)   <u>Use sweepstakes referrals.</u> Some sweepstakes give the option of multiplying your entries by telling friends and family members about the giveaway. This scheme has worked successfully for a lot of people, but I have personally never used it.

(13)   <u>Of course, prioritize your sweepstakes choices.</u>

Remember: There are far more sweepstakes available for you to enter than the time you have to enter them. Your time is limited, use it wisely.

(14)   <u>Never count on winning any one sweepstakes.</u>

I guarantee you disappointment if you do so. Enjoy the chase. Enjoy the process. Enjoy the sweepstakes regardless of whether you win. Or, don't do it.

(15)   <u>Stay motivated.</u> Keep a positive attitude. If you can do it playing golf or cards, you can do it entering sweepstakes. Optimism always helps.

(16)   Improve your luck. As we have outlined earlier in this book, you can actually accomplish this (see, Chapter V, supra).

[also see, *inter alia*, "The Psychology and Philosophy of Luck", Duncan Pritchard and Matthew Smith, Elsevier, Liverpool Hope University, England (2004); "Chance, Skill and Luck: The Psychology of Guessing and Gambling," J. Cohen, Pelican Books, London (1960); "The Belief in Good Luck Scale", P.R. Darke and J.L Freedman, Journal of Research in Personality, 31 (1997); "Lucky Events on Risk-Taking", Personality and Social Psychology Bulleting, 23, P.R. Darke & J.L. Freedman (1997); "Epistemic Luck and the Purely Epistemic", American Philosophical Quarterly, 21, R. Foley (1984)]

(17)   Make it easier for sweepstakes sponsors to contact you.

I personally subscribe to at least 2 sweepstakes publications. I know of others who check for sweepstakes on the Internet. There are several ways to do this.

(18)   Learn to recognize sweepstakes SCAMS.

By reading this book you will learn to recognize them instantaneously. And, you will know the legitimate sweepstakes. I guarantee it.

(19)   Respond to your wins promptly. They will come. And they will have time constraints attached within which you must respond: usually with an executed (signed and notarized) *Affidavit*.

(20)   Track your wins. For the following reasons,
    track your wins:
        (a) For motivational purposes. You are
        winning!
        (b) For tax reasons. Yes – you must pay
        taxes on your prizes.
        (see, "Watch our for the Tax Man", be-
        low) (c) For "win tracking" recognition
        reasons. To check for patterns and meth-
        odologies in your wins. So that you can in-
        crease your wins!

(21)   Join the Sweepstakes Community. Join
    sweepstakes clubs and other sweepstakes
    groups.

Us sweepstakers are nice, friendly and helpful people. We want you to win, too.

There are sweepstakes clubs and groups all over the
country. And sweepstakes conventions, too. Check
them out. Go on-line and search for them.

(22)   Be a considerate sweeper or sweepstaker.

Send "thank you" notes to sponsors when you win
their sweepstakes, etc.

(23)   Get inspired. Stay inspired.

Be encouraged by other people's wins. Your time
will come, too. I assure you.

(24)   Enjoy sweepstaking.

If you do not enjoy it, do not enter sweepstakes.
This is supposed to be fun… and relaxing.

(25)   Never brag or boast about your wins.

It creates bad karma. Sweepstaking is all about good karma.

In this regard I am reminded of the true but comical story of heavyweight boxer, Chuck "the Bayonne Bleeder" Wepner:

"The Bayonne Bleeder"

In 1975 heavyweight boxer Chuck Wepner, affectionately known as the "Bayonne Bleeder" because of his relatively easy propensity to bleed in his boxing matches, received a chance of a lifetime. Wepner was selected to fight the great Muhammad Ali, probably the greatest heavyweight boxer of all time, in a 15-round boxing match.

Ali was the heavyweight boxing champion of the world.

Wepner, a native of Bayonne, New Jersey, had a reputation for being a tough journeyman boxer, and he was credited to be the inspiration for actor/writer Sylvester Stallone's hit movie, "Rocky".

He bragged he was going to be the new-heavyweight champion of the world.

On the night of the big fight with the great Muhammad Ali, Wepner reports that he told his wife the following:

"Honey, the next time you go to bed it will be with the heavyweight boxing champion of the world!"

Wepner then proceeded to the boxing ring and, in the 15th round, was knocked out by Ali.

Upon returning to his hotel room to be consoled by his wife, Wepner was met with the following question by his bride:

"Should I go to Ali's room --- or is he coming to mine?"

Be humble and poised. Do not create bad karma.

Watch out for the Tax Man

One day, at a casino buffet, a man suddenly called out, "Help! My son is choking! He swallowed a quarter! Some one, please! Anyone! Help!"

A man in a suit from a nearby table stood up immediately and announced that he was quite experienced at this sort of thing.

The man stepped over to the other man's son with almost no look of concern at all, wrapped his hands around the boy's testicles, and squeezed.

Suddenly, out popped the quarter.

The good Samaritan then went back to his table as though nothing had really happened.

"My God! Thank you! Thank you so much for saving my son!" the father cried as he rushed over to the mysterious good Samaritan.

"Are you a doctor or paramedic?" he asked the stranger.

"No," replied the man. "I'm an IRS agent."

*****

America is the land of opportunity. Every lottery player, sweepstaker and contest entrant can become a taxpayer.

Sweepstakes winners: Do not forget to pay your taxes.

C. Twelve Steps For Dealing With Sweepstakes Taxes

(1) Always remember and never forget: If you live in the United States, the Internal Revenue Service (IRS) wants its share of your lottery, sweepstakes and contest wins. Consult a tax professional.

(2) Even non-monetary sweepstakes and contest prizes such as televisions, cars, etc., are classified for tax purposes as "Other Income", and, consequently, taxes must be paid on them.

(3) Keep good records of your wins and your expenses for entering and/or playing. Keep a ledger or file to track all of your entries and wins. This helps your tax preparation a lot.

(4) Do not worry excessively about your possible or probable tax ramifications until you are informed that you have won the prize. You can always decline to claim the prize, if you so choose (I had to do this once).

(5) Gather your 1099 (Miscellaneous Income) IRS forms from sponsors. This is particularly advised if you have wins with prize values of more than $600.

(6) Check the fair market value of your prizes. Your sweepstakes and contest taxes are based upon the "fair market value" (FMV), not the sponsor's estimated retail value.

(7) Total the value of your prizes. Note that <u>all</u> prizes, large and small, are legally required to be reported on your U.S. taxes.

(8) On your tax filings (1040, etc.) enter the prize total under "Other Income". This is normally line 21 of your 1040 form.

(9) Itemize your expenses. Speak with your tax consultant or advisor about this.

(10) Consult with a tax professional during this process. Sweepstakes and contest taxes can complicate a tax return. Be sure.

(11) Submit your sweepstakes information with your regular taxes. This is usually with your 1040 form.

(12) Pay your taxes! After you have sought competent tax advice from a tax professional.

[<u>And see</u>, contests.about.com/od/taxesfinances/ht/sweepstakestax.htm]

D. <u>Winning Sweepstakes On-Line: 15 Steps</u>.

(1) <u>Follow the 4 "P's"</u>:
* Patience
* Persistence
* Poise (Do not panic!)
* Positive Attitude.

(2) <u>Get a reliable computer system and connection</u>.

Your computer certainly need not be new. Just reliable. A good used computer will be fine.

(3) <u>Get a good form-filling computer program</u>.

It can make the difference between being able to do 15 on-line sweepstakes an hour and, more preferably, 100 sweepstakes an hour.

(4) <u>Organize yourself</u>.

Max-out on your entries per sweepstakes, <u>but</u> always follow the rules. Do not get disqualified by missing deadlines or submitting too many entries. Be disciplined and be organized.

(5) <u>Always read and follow the rules carefully for each sweepstakes</u>.

Not all sweepstakes rules on-line are the same, although many are similar. You do not want to waste your valuable time and effort by entering a sweepstakes you cannot win because of disqualification.

(6) <u>Be budget-minded</u> with your time, money and effort. Entering sweepstakes should be a hobby, not a job. It should be fun, not work. Be sensible and rational about these games.

(7) <u>Avoid spam as much as possible</u>.

Spam is a serious distraction for a sweepstaker. Do what you can to curtail it.

(8) <u>Avoid scams at all costs</u>.

And there are a lot of them out there. As you develop as a sweepstaker you will learn to easily spot scams.

(9) <u>Prioritize your interests and focus on deciding what contests to enter.</u>

Make sure the on-line sweepstakes you enter are able to pass your personal cost-benefit analysis and evaluation. Do not waste time and effort on sweepstakes you are not interested in, <u>for whatever reason.</u>

(10) <u>If you are sufficiently interested, from time to time take a look at the sweepstakes promoter's/company's website.</u>

You will be pleasantly surprised, sometimes, by what you see on these websites.

(11) <u>Look for, search out, on-line sweepstakes.</u>

They are all over the Internet on many different sites. On-line sweepstakers are very popular promotions for companies. There are lots, and lots, and lots of them available for you to play and enjoy!

(12) <u>WHEN YOU WIN (and you WILL WIN), properly follow the rules to claim your prize.</u>

Yes, a new set of rules arise when you win. They are simple, but you must follow them. This process is well worth it!

(13) <u>Thank the sweepstakes sponsors.</u>

Send them a "thank you" email or note. Most of them really are nice people who actually and genuinely want you to win. We appreciate their promotions, and the opportunity to participate in them.

(14) <u>When you WIN (and you WILL WIN), do not forget to pay your taxes on your prizes.</u>

Your prizes are considered "income" for IRS purposes. When you win (and you will win) you will receive the equivalent of a 1099 from the sweepstakes sponsor.

(15) <u>Enjoy yourself. Have fun.</u>

Sweepstakes are designed for fun, not for work. Enjoy yourself and you will win. I assure you.

TRACK YOUR WINS SPREADSHEET

A	B	C
	Number of Wins	**Value of Wins ($)**
January		
February		
March		
April		
May		
June		
July		
August		
September		
October		

November

December

Totals for the Year

Average Totals

Highest Month

Lowest Month

Highest Win

Lowest Win

Total Prize Count

Blog

Facebook

Instant Win Game

Internet

Local

Mail-In

Radio

Text

Twitter

Other

Prize Count Total

E. Eight (8) Reasons to Join a Sweepstakes Club.

Joining a good sweepstakes club can be very helpful to a sweepstaker for the following 8 reasons:

(1) Simple CAMARADERIE. Camaraderie refers to goodwill and lighthearted rapport and support between or among friends. A spirit of friendly good-fellowship [Merriam-Webster Dictionary].

(2) Efficient Budgeting. Group purchasing of sweepstakes (and contests, as well) supplies, equipment, etc., can be enormously helpful to the sweepstaker.

(3) Motivation and Encouragement. Fellow sweepstakers can be a marvelous source of motivation and encouragement. Members share their stories and experiences (their wins and their losses), congratulations for your wins, and encouragement during your losses (the "dry spells").

(4) Sharing News About Local Sweepstakes.

Because their number of entries received is so much lower than national sweepstakes, local sweepstakes have some of the best chances for you to win. But, hearing and finding out about local sweepstakes can frequently be tough to do. Sweepstakes club members can (and do) pool their knowledge to help locate good local sweeps in time.

(5) Swapping Entry Forms.

To enter many local sweepstakes you need to send your entry information on an official entry blank (OEB). If you don't happen to visit the location that is offering the sweepstake, you won't be able to enter.

However, sweepstakes club members will pick up extra entry forms and bring them to meetings so that their fellow club members can fill them out. This helps you enter sweepstakes that you would never know about otherwise.

(6) Sharing Sweepstakes (and Contests) Tips, Secrets and Strategies.

Is there anything about your sweeping strategy that could be improved? Talking to other knowledgeable sweepstakers can help you find out. Learn tips from organizational strategies to new ways to avoid disqualification and win more regularly at every meeting of your club.

(7) Club Group Giveaways.

Any get-together that features people who love sweepstakes is likely to have its own giveaways. Some clubs use low annual membership fees to pitch in for door prizes, while others will ask members to bring a few stamps or envelopes to participate in a drawing for sweepstakes supplies.

(8) Club Group Exchanges.

It's also very common that members will bring unwanted prize wins to swap or give away at every meeting. A win that disappointed a fellow member could be exactly what you wanted to win!

F. Publishers Clearing House (PCH) and Readers Digest Scam Prevention Advice and Tips [also see, Chapter IX, infra]

Simply because these two sweepstakes (and a couple of others, maybe) are so large and well-advertised to the public, numerous scams come along every year at the same time these sweepstakes are underway and take advantage of the consumer

[see, Legal Consumer Tips and Secrets, Charles Jerome Ware, iUniverse Publishers (2011)].

The following tips and advice should be helpful to the consumer in differentiating between these so-called legitimate sweepstakes and the many scams which attempt to take advantage of the hoopla or excitement these sweepstakes generate in the public.

Tip 1: Be aware of and cautious about fake check scams.

Advice 1: If you receive a check claiming to be from a legitimate sweepstakes and you are asked to cash it and wire or send a part or portion of it back, this is a SCAM. Stop! Do not send money. The check is a fraud. Truly legitimate sweepstakes never require or even invite a payment, fee or gift from the winner to enter or claim a prize.

Tip 2: Be very suspicious of callers, emailers, or letters from anyone claiming you have won a sweepstakes prize but insist that you send money or anything of value to claim your prize.

Advice 2: Again, STOP! This is a SCAM, not a legitimate sweepstakes win.

Tip 3: Never ever give your credit card number to claim or collect a prize.

> Advice 3: STOP! Legitimate sweepstakes do not require that you give your credit card information to collect or claim a prize. This is a scam.

Tip 4: Again, do not send money or anything else of real monetary value to claim, collect or receive your prize.

> Advice 4: STOP! This is a scam. Remember: Never pay money to an "employer" for a job --- and never pay money to a "sweepstakes" to claim a prize.

[see, Legal Consumer Tips and Secrets, Chapter Six (Employment Scams), Charles Jerome Ware, iUniverse (2011)]

Tip 5: Be cautious. Read and know the rules of the alleged sweepstakes. Know the company.

> Advice 5: A sweepstakes with no written rules is not a legitimate sweepstakes. Learn about the company sponsoring the sweepstakes.

Tip 6: Of course, if the offer sounds too good to be true, it may well be too good to be true.

> Advice 6: Measure twice, cut once. In other words, think at least twice about the sweepstakes offer, and get more information about it, before responding.

The following is an example of a phony "Publishers Clearing House" unsolicited email designed solely to scam you out of your personal information:

From: "PUBLISHERS CLEARING HOUSE" <u>info@quatec.com.br</u>
Sent: Tuesday, February 21, 2012 12:58 PM
Subject: winning numbers:47-14-34-85-67-32
We are please to announce to you that your email address emerged along side 4 others as a category 2 winner of ($1,000,000.00 USD) in this year\'s weekly Publishers Clearing House Consequently.Please contact the underlined claims officer with details below.

CONTACT EMAIL: <u>onlinepch_407@xnmsn.com</u>

1. Name in full:
2. Address:
3. Sex:
4. Nationality:
5. Age:
6. Present Country:

!!!Once Again Congratulations!!!

Yours Sincerely,
Mrs.Reyna Cruz
ONLINE CO-ORDINATOR

THIS IS A SCAM!

CHAPTER 9
HOW TO WIN CONTESTS 101

Best Video Contest

A hard-luck videographer entered a Best Video contest and won first-place. The prize was $100,000 cash money.

Being unaccustomed to having so much money, and not trusting banks, the man decided to bury the cash in his backyard.

The next day the contest winner walked outside to his backyard and, to his shock and dismay, found an empty hole where he had buried his hard-earned cash. He noticed footsteps leading from the hole to the house next door, which was owned by a deaf-mute.

Down the street lived a professor who was fluent in sign language and who was a friend of the deaf neighbor.

Grabbing his pistol, the enraged videographer rushed to the house of the professor and coerced him into going with him to the deaf neighbor's house.

"You tell this guy that if he doesn't give me back my $100,000 I'm going to kill him!" the contest winning videographer screamed to the professor.

The professor promptly conveyed the message to his friend the deaf-mute, who replied in sign language, "I hid it in my flower garden, underneath the white rose bush."

The professor then turned back to the contest winner with the gun and said, "He's not going to tell you. He said he'd rather die first."

[www.onlinecasinomansion.com/gambling-jokes.html]

It is important for the reader to know that *contests* are different from *sweepstakes*. Some skill is required for contests. Skill is not necessarily a pre-requisite for sweepstakes.

Frequently, when a sponsor offers a contest, it will require a proof of purchase. No proof of purchase is ever required for a sweepstakes. Sweepstakes are considered random drawings. Contests are definitely not random drawings.

With contests, winners are frequently chosen (or, some say, always chosen) by judges for the contestant's skill. Some of the more common contests are for photographs, videos, recipes, writings, drawings, etc. If you are skilled in an area of expertise, winning contests can be easy. Further, there are

usually far fewer contest entries to consider than sweepstakes entries.

A. 16 Tips for Winning Contests:

(1) Since contests require skill and effort, and more time
than sweepstakes, usually, decide early whether the contest
is worth putting your time, skill and effort into. Focus your
effort. Be selective.

(2) Join a contest service, club, or group, to increase your
awareness and knowledge of appropriate contests to enter.
Use multiple sources of contest listings and entries.

(3) Read and understand the rules and guidelines for each
contest you enter.

(4) Follow the rules and guidelines for each contest you enter.

(5) Follow the "4 P's":
      * Patience
      * Persistence
      * Positive Attitude, and
      * Poise (Do not panic).

(6) Organize your schedule for working on your contests.
When you set aside a specific time in your schedule, you
are more likely to do your best work.

(7) Look for contests that are less heavily publicized and/or
that may be restricted to certain age groups or geographical
regions. Search out, for example, local contests with good,
worthwhile prizes.

(8) Acquire a dedicated email address for your contest entries. Avoid using your personal and business emails. Avoid spam!

(9) Avoid aggressively advertised contests. Usually there will be too many contestants for your skill to really be recognized fairly.

(10) Highlight short entry periods for contests. If you like the contest, make the deadline.

(11) Remember: the more difficult the contest, the less competition you will have.

(12) Try to emphasize contests with multiple prizes. You want more than one (or just a few) opportunities for a prize.

(13) Budget your contest entries. Set a budget, and follow it. Do not get carried away spending a lot of money on these contests. It's not worth it. You want to enjoy yourself, not bankrupt yourself. This should be fun, not work.

(14) Do not make the mistake of believing that lower-end (less expensive) prizes are easier to win in contests and sweepstakes. This is not necessarily the case.

(15) Be careful about giving out your cellphone number when the contest sponsor <u>asks</u> for it. Frequently in these situations, it is a telemarketing ploy.

(16) Never forget – and always remember – that "contesting" (as well as "sweepstaking") is, or should be, a hobby --- not a job. Do not get carried away with it. Enjoy this hobby!

B. 6 Reasons for Entering Skill Contests:

(1) <u>Contests (by definition they require skill)</u> are lots of fun! They present opportunities for you to engage in activities in which you have skill and which you enjoy. And, you can be rewarded with prizes for your efforts!

(2) <u>Contests have less competition than sweepstakes</u> and lotteries. Generally, contests have a fraction of entries that sweepstakes and lotteries have. Therefore, you have a better chance of winning.

(3) <u>Contests are not random, or based on luck.</u> Your skills are what you count on in contests. You have contest judges you can impress with your skills.

(4) <u>Contests provide opportunities for you to use and reveal your creativity.</u> What fun!

(5) <u>Contests provide organized opportunities for you to practice, hone, and improve your skills.</u>

Contests motivate you to practice and improve your skills. This way you have less chance of losing your skills. You either use them or lose them.

(6) <u>Winning contests, using your own skills, is a true achievement for you.</u>

You have accomplished something wonderful that is not based upon luck or chance. Take pride in that achievement! *Congratulations!*

<u>The "Computer Competency""Contest</u>

Jesus and Satan have an argument as to who is the better computer programmer. This goes on for a few hours until they agree to hold a "computer competency" contest with God as the judge.

They set themselves before their computers and begin. They type furiously for several hours, lines of code streaming up the screen.

Seconds before the end of the competition, a bolt of lightning strikes, taking out the electricity. Moments later, the power is restored, and God announces that the contest is over. He asks Satan to show what he has come up with. Satan is visibly upset, and cries, "I have nothing! I lost it all when the power went out."

"Very well, then," says God, "let us see if Jesus fared any better."

Jesus enters a command, and the screen comes to life in vivid display, the voices of an angelic choir pour forth from the speakers. Satan is astonished.

He stutters, "But how?! I lost everything yet Jesus' program is intact! How did he do it?"

God chuckles, "Jesus saves."

[www.jokebuddha.com/contest]

CHAPTER 10
WINNING THE LOTTERY BY AVOIDING LOTTERY SCAMS

 ops!

At a Christmas party the staff decided to pull a joke on their boss who had a habit of playing serious practical jokes on everyone else. When he went to the toilet, they went through his wallet and found his Lotto ticket. Then, they wrote down his numbers and called over the waitress to set up a little prank.

She came back half an hour later and asked if anyone wanted to know the night's Lotto numbers, then proceeded to read them out loud before setting the numbers on the table.

The boss looked at the numbers, then casually pulled out his wallet and compared them. He became really silent, put his wallet back in his jacket and sat down again breathing really rapidly, and looking totally blown away.

After a couple of minutes he pulled out his wallet and Lotto ticket again, and checked the numbers, very carefully.

Then, he downed his drink, stood up on his chair and shouted out to the whole room:

"I just want to let you all know something. I've been having an affair with my secretary for months. I don't like any of you, and I have hated working for this company. You can all go to Hell, because I've just won a shit-load of money, and I'm leaving!"

End of job. End of marriage. End of story.

[www.lottomania.ch/humour.php]

Successful lottery winners know how to avoid lottery scams, jokes and pranks. So, be careful.

A. Ten (10) Tips for Identifying Lottery Scam Emails and Letters:

(1) If you, or someone else <u>for</u> you, did not buy a lottery ticket, you cannot win the lottery. You must play the game in order to win the game. No ticket, NO WIN.

(2) The lottery is unknown, or unidentifiable. Unless, of course, you find it on a lottery scam website. In any event, avoid it. It is most likely a scam.

(3) The lottery's name is a company name, like Gateway or Verizon. Lotteries are <u>not</u> sponsored by merchants, companies, or individuals, except through a *sweepstakes*.

(4) Again, remember ((3), above), lotteries are <u>not</u> sponsored by individuals, no matter how wealthy they are. Bill Gates or the Crown Prince of Saudi Arabia, or some official in Nigeria, do not, and cannot, sponsor lotteries.

(5) You do not live in the lottery country, and you are not a citizen of the lottery country. Most lotteries are restricted to residents of the state, province, or country in which the game is played.

(6) Since no one is permitted to use your email address without your specific and expressed permission, it is not possible for your email address to win in any random email drawing.

(7) It's a scam letter if you did not register your name, street address, email address, telephone number, and a credit card BEFORE you were allowed to buy a ticket on an online pay-to-play lottery web site.

(8) You cannot win the lottery simply by participating in surveys. Legitimate surveys do not purchase lottery tickets for you. Invariably, surveys that promise lottery tickets are probably ID Theft scams. Be aware of this.

(9) Legitimate lotteries (and sweepstakes as well) generally advertise their games on a daily basis. If your alleged lottery is not doing that, and you are not aware of them, or it, be extremely cautious.

(10) Remember: If you are required to pay money <u>after</u> you have allegedly won the lottery, it is probably a scam.

[<u>Legal Consumer Tips and Secrets</u>, Chapter 11, Charles Jerome Ware, iUniverse Pub. (2011); <u>Understanding the Law: A Primer</u>, Charles Jerome Ware, iUniverse Pub. (2008);

www.fraudaid.com/scamspam/lottery/identifylotteryscamle tters]

B. Avoid Scam Lottery Emails and Letters.

(1) Lottery scam emails and letters are sent out by multiple thousands every day to potential victims. These scam emails and letters have only three (3) purposes:

 (a) to take your hard-earned money;
 (b) to steal your identity; and
 (c) to involve you in the illegal activity of money laundering (by sending you counterfeit and otherwise fraudulent checks and money orders).

(2) Stealing your identity is just as important, and frequently more important, to these scammers than stealing your money.

(3) Identity theft is a very serious and widespread crime.

[Legal Consumer Tips and Secrets, Chapter 5, Charles Jerome Ware, iUniverse Pub. (2011)]

C. You Can Be A Victim And Still Go To Jail.

Remember: It is your legal responsibility to verify the legality of checks and money orders when you cash or deposit a draft or receive funds wired into any of your bank accounts. By law, you are legally responsible and fully liable for any checks or money orders you cash anywhere in the world. Be aware that monies wired into any of your bank accounts by a stranger may be stolen funds. Therefore, you and your accounts could be used for *money laundering.*

Nigerian, Asian and Eastern European organized crime networks, particularly, are known to specialize in sending counterfeit checks and money orders to unsuspecting scam victims for the purpose of money laundering. These same crime networks acquire the account information of their scam victims and wire money stolen from another account holder into the scam victim's account. In both cases, the scam victim is told to send a large portion of the money by Western Union or MoneyGram to fictitious names, usually in foreign countries. The scam victim has now engaged in *money laundering* and will owe the entire amount of money he or she has sent off. Further, and worse, the scam-victim is at high risk and exposure of being criminally charged, indicted, for theft, forgery, and money laundering.

D. It is Illegal for U.S. Citizens to enter Foreign Lotteries.

[see, Racketeering (Federal Statute), U.S. Code, Title 18, Part 1, Chapter 95; www4.law.cornell.edu/uscode/html/uscode18/uscsup01181 012095.html]

E. The Green Card Lottery Scam.

[see, The Immigration Paradox: Fifteen (15) Tips for Winning Immigration Cases, Charles Jerome Ware (former U.S. Immigration Judge, iUniverse Pub. (2009); Quince (15) Consejos Para Ganar Casos De Inmigracion, Charles Jerome Ware, ex Juez de Inmigracion de los Estados Unido, iUniverse (2011); U.S. State Department Website]

Tips to remember about the U.S. Green Card Lottery, generally:

(1) The State Department Website for the Diversity Visa Program (Green Card Lottery) generally opens in the Fall or Winter of each year, and usually every year.

(2) The program (U.S. Diversity Visa Lottery Program) is free. Registration is free.

(3) The program is usually open for 60 days.

(4) Applicants must complete the application form in order to win this lottery.

(5) Winners of the program are not notified by email. Winners are notified by the U.S Government, only, and by surface mail, USPS, only.

(6) Applicants do not have to (are not required) to pay anyone to assist or help them with their application.

(7) Be aware of, and avoid, scam green card lottery emails and web sites.

(8) Be advised: There is no such thing as the 2009, 2010, 2011 or any other year "Green Card Lottery". The program is generally called the Diversity Visa Lottery Program.

(9) Applicants may apply only once a year, but they may apply every year.

(10) The Diversity Visa (DV) Lottery Program Schedule is numbered by the fiscal year in which visas are issued to the qualified winners: example, 2009, 2010, 2011.

(11) Tip: The U.S. Government will assist you in reviewing and understanding the application form for free. Contact your local U.S. Embassy or Consulate for help. When in doubt, ask.

(12) For the applicants' convenience, DV Lottery instructions are available in several languages.

CHAPTER 11
BASIC LOTTERY AND SWEEPSTAKES MATHEMATICS AND STATISTICS: A PRIMER

et Lotto Ticket?

One day, the wife comes home with a spectacular diamond ring.
"Where did you get that ring?" her husband asks.
"Well," she replies, "my boss and I played the lotto and we won, so I bought it with my share of the winnings."
A week later, his wife comes home with a long shiny fur coat.
"Where did you get that coat?" her husband asks.
She replies:
"My boss and I played the lotto and we won again, so I bought it with my share of the winnings."
Another week later, his wife comes home, driving in a red Ferrari.
"Where did you get that car?" her husband asks.

Again she repeats the same story about the lotto and her share of the winnings.

That night, his wife asks him to draw her a nice warm bath while she gets undressed. When she enters the bathroom, she finds that there is barely enough water in the bath to cover the plug at the far end.

"What's this?" she asks her husband.

"Well," he replies, "we don't want to get your lotto ticket wet, do we?"

[www.lottomania.ch/humour.php]

Because of the sheer technical difficulty of the subject, this chapter on the basic mathematics and statistics of lotteries and sweepstakes is deliberately presented <u>after</u> the more practical and straight forward "how to win" chapters.

I believe that even the staunchest mathematician, when pressed, will admit the following feeling about mathematics and statistics:

"Mathematics" and "statistics" are made of 50 percent formulas, 50 percent proofs, and 50 percent imagination. ☺

[modified and amended,

www.math.utah.edu/mathjokes.html]

A. The Poisson Distribution.

The number of winners in a lottery has an approximate *Poisson Distribution*, which is always known as the *Poisson law of small numbers:* the discrete possibility distribution that expresses the probability of a given number of events occurring in a fixed interval of time and/or space if these events occur with a known average rate and independently of the time since the last event:

$$f(k;\lambda) = \lambda^k e^{-\lambda}\big/_{k!}$$

where

- e is the base of the natural logarithm ($e = 2.71828...$)
- k is the number of occurrences of an event — the probability of which is given by the function
- $k!$ is the factorial of k
- λ is a positive real number, equal to the expected number of occurrences during the given interval. For instance, if the events occur on average 4 times per minute, and one is interested in the probability of an event occurring k times in a 10 minute interval, one would use a Poisson distribution as the model with $\lambda = 10\times4 = 40$.

As a function of k, this is the probability mass function. The Poisson distribution can be derived as a limiting case of the binomial distribution.

The Poisson distribution can be applied to systems with a large number of possible events, such as lotteries and sweepstakes. The Poisson distribution is sometimes called a Poissonian.

EXAMPLE of the Poisson Law of Distribution

A store sells on the average 3 winning lotto tickets per week. Use Poisson's law to calculate the probability that in a given week the store will sell:

a. Some winning lotto tickets.
b. 2 or more tickets but less than 5 winning lotto tickets.
c. Assuming that there are 5 ticket-selling days per week, what is the probability that in a given day the store will sell one winning lotto ticket?

ANSWER

In this example, M equals 3 winning lotto tickets.
 (i) "Some tickets" means "1 or more tickets". We can work this out by finding 1 minus the "zero tickets" probability:

$P(X > 0) = 1 - P(x_0)$
- 95% is the probability the store will sell 1 or more winning lotto tickets;
- 62% is the probability the store will sell 2 but less than 5 winning lotto tickets in a 5-day week; and
- 33% is the probability the store will sell one winning lotto ticket in a given day.

TIP: Purchase your lottery tickets from the stores that have the best lottery-winning results or history!

[Mathematics from the Birth of Numbers, Jan Gullberg, W.W. Norton: New York (1997); Probabilite' des jugements en matiere civile, precedes des regles generales du calcul des probabilities, S.D. Poisson, Bachelier: Paris, France (1837)]

The expected number of lottery winners is the number of lottery tickets sold divided by the number of ways to win.

This number, for example, is approximately 146.1 million for the current Powerball Lottery.

As we discussed previously in Chapter Five ("How To Win Lotteries 101"): as the size of the jackpot increases, so does the expected number of winners; and, the expected winnings decrease because the winners have to split the jackpot; and, even for very large jackpots, the expected value of a ticket is still less than the price of the ticket --- unless and until your odds are improved by following the tips and secrets in this book.

For the mathematically inclined, a formula to predict the expected jackpot winnings (as derived by way of the *Poisson Distribution* is:

$$W = J(1 - e^{-m}) / m$$

where

- W is your expected jackpot winnings,

- J is size of the jackpot

- m is expected number of winners, and

- e is base of the natural logarithms (use the e^x key on a scientific calculator to calculate e^{-m}).

B. Calculation of the Probabilities in a Lottery Game: Lottery Mathematics.

Lottery mathematics is a term used frequently to refer to the calculation of the probabilities in a lottery game.

In our discussion here we will assume the lottery game is one in which a player selects 6 numbers from a total of 49 numbers (from 1 to 49). Let's assume this is a POWERBALL-type Lottery.

In a truly random 6/49 lottery game, six numbers are drawn from a range of 1 to 49 numbers, and if the six numbers on a ticket match the 6 numbers drawn, the ticket holder wins the lottery (the "jackpot"). This is true regardless of the order in which the six numbers appear.

The mathematical or statistical probability of this happening in a truly random drawing is *1 in 13,983,816!*

The fundamental goal, of course, for the reader of this book is to increase (↑) his or her probability of winning this lottery and beating those slim odds.

["On the Lottery Problem", Journal of Combinatorial Designs, Z. Furedi, G.J. Szekely, and Z. Zubar, Wiley Pub. (1996); "Introduction to Lottery Mathematics: Probabilities, Appearance, Repeat, Affinity or Number Affiliation, Wheels, Systems, Strategies", Ion Saliu, saliu.com/gambling-lottery-lotto/lottery-math.htm (Retrieved 01/16/2012)]

The ordinary or random chance of winning the 49-numbers lottery can be shown as follows (using the POWERBALL-type Lottery as the example):

1. There are 49 different numbers (say, 1 through 49); and there are, therefore, 49 different but equally likely ways of choosing the first number selected. That means there is a 1 in 49 chance of predicting the first number correctly.

2. Now, the <u>second number</u>. There are now 48 numbers left to choose from, so there is now a 1 in 48 chance of predicting the second number.

3. This pattern continues. For example, there are now 47 numbers remaining to choose from, so there is now a 1 in 47 chance of predicting the 3rd number. And continuing on, there would be a 1 in 43 chance of predicting the 6th number, for example.

4. This means the probability of correctly predicting 6 numbers drawn from 49 numbers in the correct order can be calculated as:

1 in 49x48x47x46x45x44x43.

This formula could be written as: $49! / (49 - 6)!$

["On the Lottery Problem", Furedi, Szekely and Zubor, <u>Journal of Combinatorial Designs</u>, (1996), <u>supra</u>]

C. Smaller Number Basic Lottery Mathematics and Statistics.

Particularly when it comes to smaller-number lotteries such as "pick 3" and "pick 4" games, so-called "hot" numbers tend to travel in what we call <u>clusters</u> (groups).

These *clusters* tend to be comprised of, for instances:

1. *A Mirror*: say, 422-224

2. *A Mate*: say, 422-533

3. *A Follower*: say, 422-424

4. *A Cluster*: say, 105-115-116-126-063-411

5. *Relatives for the Number 456*: 906, 294, 005, 447, 688, 159, 258, 339, 366, 555, etc.

Visualize (see) the number patterns!

[Also see, supra, Chart on "Winning Pick-3 and Pick-4 Lottery Games", Chapter 3, How To Win Lotteries 101]

D. Hypergeometric Distribution and the "Bernoulli Trial": Figuring the Odds and Probabilities.

Much odds and probability theory is based on the notion of a "Bernoulli trial" -- an event that results in a "success" or "failure." When tossing a coin, for example, you can consider "heads" a success and "tails" a failure, each occurring with a probability of 1/2. When attempting to calculate the probability of a sequence of coin tosses -- say, the chance of getting two heads in a row -- you only need to multiply the probability of each individual event. For two heads in a row, the probability would be 1/2 x 1/2, or 1/4. Odds for the "Pick 3" game can be calculated in this way; the general formula is called the "binomial distribution."

However, for 6-number lotto games such as Powerball, Mega Millions, and many state lotteries, the binomial distribution does not apply. When tossing coins, the result of one toss does not affect the result of the next. The events are said to be "independent." But, in a lotto game, selection of successive balls is not independent because the balls are not put back into the machine. For a 59-number Powerball, the probability of the first ball being a 1 are 1 in 59. If a 1 is selected, the probability of a 1 on the second ball is 0. If 1 is not selected on the first ball, the probability of selecting it on the second ball drops to 1 in 58 because there is one fewer ball in the machine. This is referred to as "sampling without replacement."

[See, Chapter V under "Hypergeometric Distribution", and "Winning 'Pick 3' and 'Pick 4' Lottery Games'", as well as Chapter II, "Odds and Probabilities" and "The Concept of 'Independent Events'", *supra*]

EXAMPLE

For example, assume we have F objects from which to choose. (For Powerball, this would be the 59 white balls.) Of these F, M are successes (numbers that match the ones on your ticket, or 5 in the case of Powerball) and F-M are failures (for Powerball, the 54 balls in the machine that don't match a number on your ticket).

Next we conduct n Bernoulli trials - we draw n balls (5 for Powerball and Mega Millions) from the machine. What we need to know is the probability of getting p successes and n-p failures in those n trials. To match all five white balls, for example, you calculate the probability of 5 successes and 0 failures in 5 trials. The general formula for probability is (Equation 1):

$$\frac{\text{(number of ways of getting p successes) x (number of ways of getting n-p failures)}}{\text{(total number of ways of selecting objects from a set of F)}}$$

Each of these terms boils down to counting.

To calculate the denominator, begin by realizing that there are F ways of selecting the first object; F-1, ways of selecting the second object; and so on to F-n ways of selecting the nth object. The total number of ways of making this selection, therefore, is $F(F$-1$)(F$-2$)...(F$-$n)$. However, in our case the order of selection does not matter -- (1,2,3,4,5) is the same as (5,4,3,2,1). We need to adjust for the number of combinations that are identical except for the order.

In order to visualize this, imagine a drawing of two balls from a set of three. There are 3 ways of picking the first ball and two ways of picking the second, for a total of 6 outcomes: (1,2), (1,3), (2,1), (2,3), (3,1), and (3,2). However, only three of these are distinct: (1,2), (1,3), and (2,3) -- the others are merely reorderings.

In general, if we are picking n objects, there will be $(n)(n-1)(n-2)$... ways of arranging each unique combination, so we need to divide our first calculation by this term. There's a mathematical shorthand for this, called a "binomial coefficient."

[See for support, "An Introduction to Probability Theory and Its Applications", Volume 1, William Feller, John Wiley & Sons, Inc.: New York (1968); www.lottery.state.mn.us/hypergeo.html]

Still, It Takes More Than Brains To Win

"Congratulations!!!" were showered on Martha.

Her number 49 had won the top prize in the lottery.

"Wow, say Martha", asked Jerome, "how did you happen to pick lucky number 49?"

"I saw it in a dream. Six sevens appeared and danced magically before my eyes. Six times seven is 49, and that's all there was to it," answered Martha.

"But Martha," Jerome responded, "six times seven is 42, not 49."

"Alright already," Martha continued, "so you be the mathematician and I'll stay the winner!"

CHAPTER 12
A SUMMARY OF THE SECRET SCIENCE OF WINNING LOTTERIES, SWEEP- STAKES AND CONTESTS

Luck Counts but Preparation Rules!

Neil Alden Armstrong --- former American astronaut, test pilot, aerospace engineer, university professor, United States naval aviator, and the first person ever to set foot on the Moon --- has always epitomized the traits of persistence, preparation, poise and positive attitude.

As mission commander for the Apollo II moon landing mission on July 20, 1969, when Armstrong walked on the Moon, he not only gave his famous "One small step for man, one giant leap for mankind" statement, he followed it with several additional remarks.

One of the additional remarks he made as he successfully stepped back onto the landing craft from the Moon's surface was the enigmatic remark, *"Good luck Mr. Gorsky!"*

Many people at NASA thought it was a casual remark concerning some rival Soviet Cosmonaut. However, upon checking, there was no Gorsky in either the Russian or American Space Programs.

Over the years many people questioned Mr. Armstrong as to what the *"Good Luck Mr. Gorsky"* statement meant, but Mr. Armstrong always just smiled and would not answer.

On July 5, 1996, in Tampa, Florida while answering questions following a speech, a reporter brought up the 29 year old question to Mr. Armstrong again. This time he finally responded.

Mr. Gorsky had finally died and so Neil Armstrong felt he could answer the question:

When he was a kid, he was playing baseball with a friend in his backyard. His friend hit a fly ball which landed in the front of his neighbor's bedroom windows. His neighbors were Mr. and Mrs. Gorsky. As he leaned down to pick up the ball, the then young Neil Armstrong heard Mrs. Gorsky shouting at Mr. Gorsky saying, "Oral Sex! You want oral sex? You'll get oral sex when the kid next door walks on the Moon!"

Statistical analysis reveals that the odds or probability at that time that young Neil Armstrong of Wapakoneta, Ohio would one day walk on the Moon were from 12,100,000 to 1 up to incalculable!

It is believed that --- thanks to Armstrong's persistence, preparation, poise and positive attitude, which resulted in him actually walking on the Moon

--- his former neighbor Mr. Gorsky was able to get lucky. ☺

[www.jokebuddha.com/Luck (as modified)]

There is a science of winning lotteries, sweepstakes and contests. This book explains this science with some humor and a lot of factual detail.

Blind reliance on luck or chance is not necessary to win. Persistence, preparation, poise, positive attitude, strategic planning and cognitive flexibility learning are necessary for any player to win on a consistent or regular basis. Therefore, just about anyone can win.

When it comes to lotteries, sweepstakes and contests, there are ways to improve your odds or probability of winning. They are discussed in this book in detail, and they work. If you have studied this book, you can win.

Congratulations! You have won!

POSTSCRIPT: Now --- What To Do After You Win The Lottery, The Sweepstakes, Or The Contest

---

1. If you have won the Lottery, make sure you have <u>signed</u> the back of your lottery ticket in the proper place in ink.

2. If you have won the Sweepstakes or the Contest, make sure you properly <u>execute</u> the Affidavit to receive your prize in a timely manner.

3. <u>Safekeeping</u>. Protect that lottery ticket at all times. The moment you become aware or truly believe you have won the Lottery, get a safe deposit box at the bank or acquire a very reliable home safe and place your winning lottery ticket in it. This way your ticket will be protected from loss or theft until you can turn it in to the proper lottery officials.

4. <u>Lump sum or installment payments</u>. If the option exists, you will need to decide whether to choose a single lump sum payment (minus taxes, of course) of your lottery win or installment payments (minus taxes).

   With sweepstakes and contest wins you, generally, will receive a form 1099 for your win at the end of the taxable year.

5. <u>Hire Professionals</u>. It is important as part of your decision-making processes to hire capable, experienced, and trustworthy accounting, tax, legal, and financial advisors to help you.

6. <u>Privacy</u>. Consider the ramifications of <u>who</u>, if anyone at all outside of your advisors (and family), you want to know about your win.

   Never forget that your life will change drastically and forever, and in many ways, when you win a large prize of any type (and especially when you win the lottery).

   You will be inundated and targeted by numerous scammers and schemers, thieves and other crooks, for-profit and non-profit entities, long-forgotten sweethearts, so-called new friends, etc., seeking some or all of your new wealth. Establish an investment plan first and early before dealing with these people.

Generally, in these instances, silence is golden.

7. <u>Bank Accounts</u>. Make sure your selected bank or other financial institution is aware of your big money win, and that your account for the money is set up properly to immediately receive the transfer or deposit of funds.

8. <u>Financial Advice</u>. Again (see 5, *supra*), make sure you have a good financial advisor to help you make your financial decisions.

9. <u>Follow the Rules</u>. Again, follow all of the established rules and procedures, and timelines to properly acquire your prize.

10. <u>Do not quit your job… just yet!</u>

Nor should you "act out" in any way. Be careful! Maintain (or acquire) good karma. Make very sure you have actually <u>won</u> before making any life-altering changes and decisions such as quitting your job. Maintain (or develop immediately) poise! Do not make the mistake of the "boss at the Christmas party" [see, "Oops!", Chapter X: Winning The Lottery By Avoiding Lottery Scams, *supra*]. ☺

Good luck!

INDEX

Chapter III: The 4 P's

Chapter IV: Lotteries in the United States

Chapter V: How To Win Lotteries 101: Cognitive Flexibility Learning

Chapter VI: How To Win Lotteries 102: Winning Lottery Number Formulas, Patterns and Techniques

Chapter VII: How To Create Your Own Winning Lottery Algorithm

Chapter VIII: How To Win Sweepstakes 101

Chapter IX: How To Win Contests 101

Chapter X: Winning the Lottery By Avoiding Lottery Scams

Chapter XI: Basic Lottery and Sweepstakes Mathematics and Statistics: A Primer

Chapter XII: A Summary of the Secret Science of Winning Lotteries, Sweepstakes and Contests

CPSIA information can be obtained
at www.ICGtesting.com
Printed in the USA
LVOW03s0914290317
528891LV00001B/80/P